ELEMENTARY ENGINEE

Other title by the same author
Applied Mechanics Made Simple (Heinemann: 'Made Simple' series)

Other Macmillan titles of related interest
Basic Engineering Mechanics J. H. Hughes and K. F. Martin
Engineering Mathematics, second edition K. A. Stroud
Further Engineering Mathematics K. A. Stroud

Elementary Engineering Mechanics

G. E. Drabble

First published 1986
Reprinted 1990

Published by
MACMILLAN EDUCATION LTD
Houndmills, Basingstoke, Hampshire RG21 2XS
and London
Companies and representatives
throughout the world

Printed in Hong Kong

British Library Cataloguing in Publication Data

Drabble, G. E.
Elementary engineering mechanics.
1. Mechanics, Applied
I. Title
531'.02462 TA350

ISBN 0–333–39926–9
ISBN 0–333–39927–7 Pbk

Contents

Introduction

This book comprises a set of three 'programmed texts' designed to help you in learning the fundamental principles of statics, kinematics and kinetics. You may have encountered similar texts elsewhere, but if you have not, you should understand that they are designed to be read, one Frame at a time, in order, without skipping, and without looking ahead at subsequent Frames out of order. Moreover, you should not attempt to work through Programme 3 (Kinetics) until you have thoroughly covered the work of Programmes 1 and 2 (Statics and Kinematics). When working through the texts you might find it helpful to have a card or a sheet of paper, to cover up the Frames ahead of you, so that you do not see answers to any of the questions asked before you have had time to work out the answer yourself. Wherever questions are asked, or exercises set, it is very important that you attempt to answer the questions, or go through the exercises, before reading on; this is how you will learn. If you make any mistakes (and you almost certainly will) you should hopefully find out your errors from the subsequent Frames. Try to understand clearly the reason for your mistake. If you cannot do this, then make a note, and get some help from your lecturer as soon as you get the opportunity.

To work through these texts, you will need

1. A ruler, or draughtsman's scale
2. A protractor
3. A set-square
4. A 2H (or equivalent) pencil
5. A pen and paper
6. A calculator.

You should get guidance from your lecturers about when you should read these texts, but you should understand that they are designed as revision for A-level or equivalent work, so that ideally, you should have completed working through them before you start on the first year of a Degree course in Engineering. Like most students, you have probably been accepted for a Degree course while feeling that you did not perform as well as you had hoped on your A-levels. You may have been lucky enough to scrape a Grade C in Physics or Applied Mathematics and still have only a very vague notion of centroids, or moments of inertia. Here is a chance to catch up.

Programme 1: Elementary Statics

1

The study of Engineering Mechanics is largely concerned with examining forces and the effects that they produce. One important effect of force is to produce an acceleration. This aspect of force will be examined in programme 3, called 'Elementary Kinetics'. Sir Isaac Newton showed that force is a phenomenon which causes, or tends to cause, acceleration of a body. These words, 'tend to cause', need explanation. Why is it that forces sometimes cause acceleration and sometimes not? To help answer this question, let's look at a body subjected to two very simple, but different force situations.

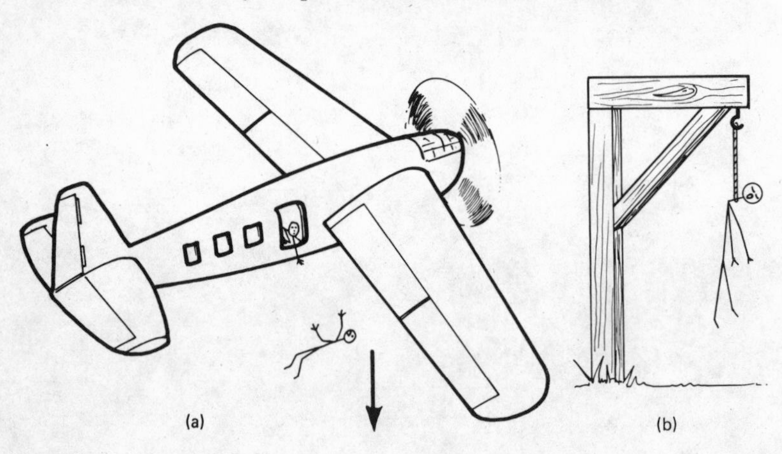

In sketch (a) a man has just fallen from an aeroplane. It is clear that while he is falling, he is accelerating; that is, his speed of fall is increasing. The force causing this acceleration is his weight—the pull of the earth on his body. It must be this force, because there are only two forces acting on his body: his weight, and the resistance of the air. The air resistance acts in an upward direction, opposite to the direction of motion, and has the effect not of actually slowing him down, but of slowing down the rate at which his speed of fall is increasing—his acceleration, in other words: his acceleration gets less the further he falls. So, in this example, his weight causes his body to accelerate downwards.

Now look at sketch (b). The same man (who, we have to admit, is rather unlucky) is now hanging from a gallows. His weight still acts upon his body (no doubt much to his regret). But this time, it is not causing his body to accelerate. This reason is clear; his body is being pulled *down* by the pull of the earth, but also it is being pulled *up* by the rope around his neck. And the two pulls are equal and opposite.

Forces, then, produce an acceleration on a body only when they are not balanced out, or cancelled, by other forces. In such cases we say that a *resultant* force exists. The determination of the effects of a resultant force on the motion of a body forms the study of *Kinetics*, and this is also dealt with in Programme 3, 'Elementary Kinetics'. The foundation-stone of Kinetics is the three Laws of Motion established by Sir Isaac Newton in the seventeenth century. The first of these laws states that if a body is not subjected to a resultant force, it will be at rest, or in a state of uniform straight-line motion. The word 'static' means 'at rest', and Statics began as the study of forces acting on bodies at rest, such as buildings and bridges. But the word has now come to have a wider meaning, and you should think of Statics as the general study of force systems. As such, it becomes an essential prerequisite of Kinetics.

2

We can begin with the simplest possible force system which can be in equilibrium. Obviously, a single force cannot be in equilibrium; a body subjected to a single force must accelerate. But two forces can be in equilibrium. Our friend of the previous Frame is in equilibrium, hanging from the gallows. (We are, of course, referring to the forces acting on his body, and not to his mental state.) So make use of him as an example, and deduce what conditions are necessary in order that two forces shall be in equilibrium. Before you plunge in with a hasty answer, let me remind you that *three* conditions are necessary.

3

The conditions for two forces to be in equilibrium are

1. The two forces must have the same magnitude
2. They must act in opposite directions
3. They must act along the same straight line—that is, they must be collinear.

Now think of some examples of two-force systems which conform to two of these conditions but not to all three.

4

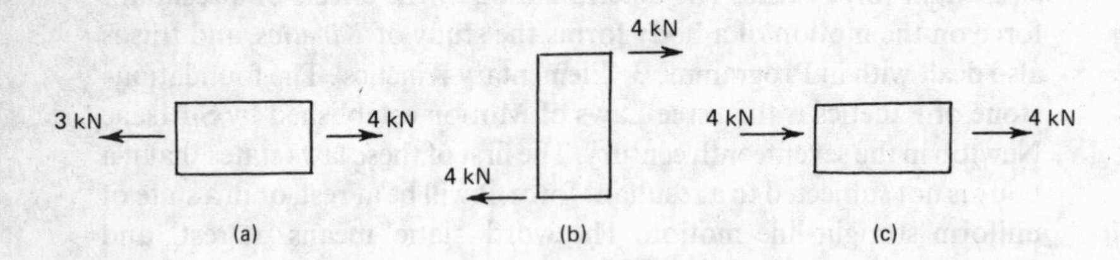

In example (a) the forces are opposite and collinear but not equal. In (b) they are equal and opposite but not collinear. In (c) they are equal and collinear but not opposite in direction. In none of these cases, therefore, does equilibrium exist; the *three* conditions are all required. You might be tempted to think that (b) represents equilibrium, but you can probably see that if you applied a pair of equal forces to a body in this manner, it would turn. (This is one way that a pair of tugs might turn a liner in dock.) Such an arrangement of equal parallel forces is called a *couple*.

Before we can consider the conditions for the equilibrium of three or more forces acting at a point, we have to state a very important principle of statics known as the Parallelogram Law, or the law of composition of forces. This law was formulated by Stevinus in the sixteenth century. We shall state the law without proof, although it is probable that in your previous work in Mechanics you have encountered an experimental proof or verification of the law. The law is stated in Frame 5.

5

The Parallelogram Law states

"If two forces acting at a point O are represented by lines OA and OB, where OA and OB are proportional in length to the magnitudes of the forces, and their directions are those of the forces, the effect of the two forces is the same as that of a single force represented in magnitude and direction by a line OC, where OC forms the diagonal of the parallelogram of which OA and OB form adjacent sides"

In the example following, a body is subjected to forces of 2 kN and 3 kN in the directions shown. The effect of these forces is the same as that of a single force R, which is determined by drawing the parallelogram and measuring the length of the diagonal. OA is drawn 2 units long to represent 2 kN, in the direction of this force, and similarly, OB is made 3 units long in the corresponding direction. OC is the resulting diagonal of the parallelogram produced by drawing AC and BC parallel to OB and OA respectively. In this example, the length of OC scales approximately 4.36 kN and the angle α measures approximately 37°C. The single force R represented by vector OC is called the *resultant* of the two forces.

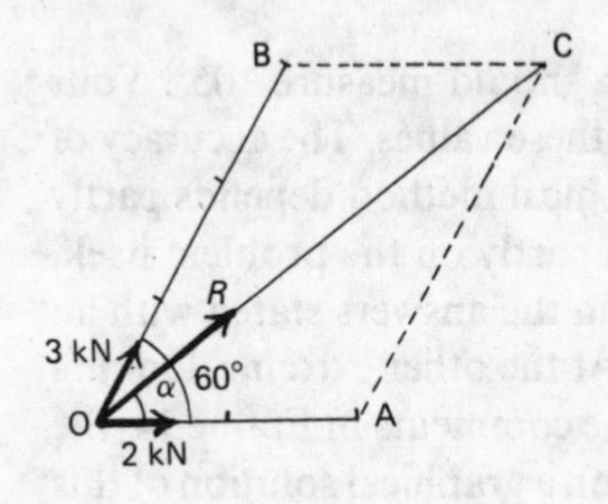

OA measures 2 units
OB measures 3 units
OC scales 4.36 units

Now make use of this same principle to find the resultant of the following two forces

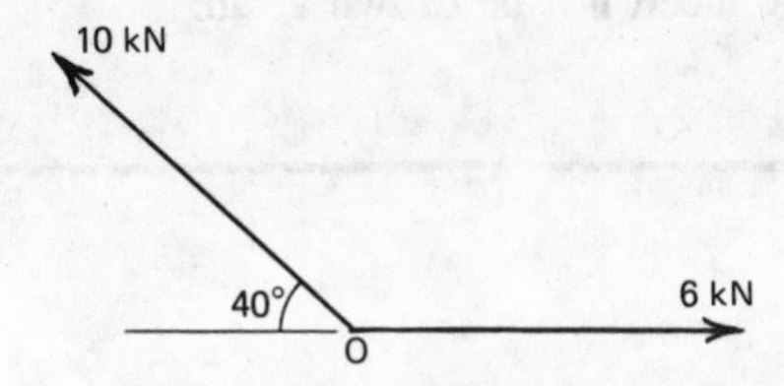

and check your work against the solution in the Frame following.

6

The parallelogram, drawn to scale, should look like this

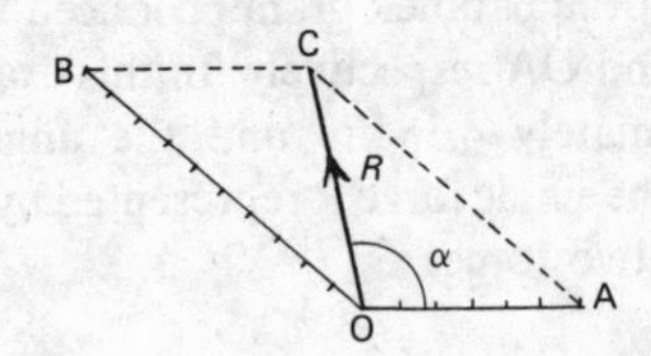

OC should scale 6.63 kN and the angle α should measure 105°. Your own solution may not agree exactly with these values. The accuracy of the answer to a problem solved by a graphical method depends partly upon the skill and care of the solver, and partly on the problem itself. Taking reasonable care, you should obtain the answers stated with an error of not more than about 1 per cent. At the other extreme, looking ahead to the problem of Frame 12, and the comments in Frame 14, the same degree of skill and care expended upon a graphical solution of this problem could result in a percentage error very much greater than this.

Returning to the problem of this present frame, you may have noticed that you did not need to draw the complete parallelogram. The length and direction of OC can be found simply by completing the triangle OAC. BC need not be drawn at all.

7

Again using exactly the same technique, find the resultant of these two forces

8 kN

90°

6 kN

Just draw the triangle, or 'half-parallelogram' OAC; do not bother to draw BC.

8

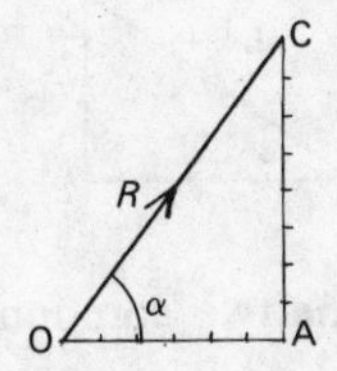

Completing the triangle OAC gives OC = 10 kN at $\alpha = 53°$ to the direction of the 6 kN force. In this case, the values can be *calculated* from the trigonometry of the right-angled triangle, thus

$$R = \sqrt{6^2 + 8^2} = 10\ \text{kN}$$

$$\alpha = \tan^{-1}\left(\frac{8}{6}\right) = 53.1°$$

This example serves to introduce the principle of *Resolution of Forces*. The Parallelogram Law enables us to determine a single force which replaces two others. Resolution is just the opposite; a single force is replaced by two *components*. The reason for doing this will appear shortly. When forces are resolved in this manner, we always calculate two components which are at right-angles to each other. As a first exercise, find the two components of the 45 kN force shown below, along the two directions OX and OY.

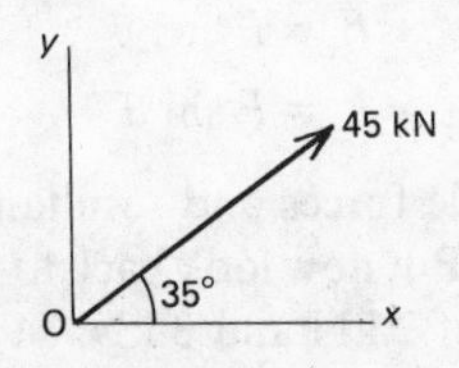

9

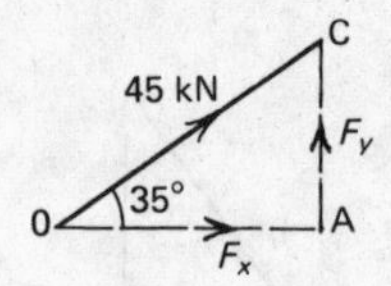

45 kN is the resultant of the two components F_x and F_y. From the triangle OAC

$$F_x = 45 \cos 35° = \underline{36.86 \text{ kN}}$$

$$F_y = 45 \sin 35° = \underline{25.81 \text{ kN}}$$

Now repeat this exercise, except that 45 kN is replaced by F, and 35° is replaced by θ, thus

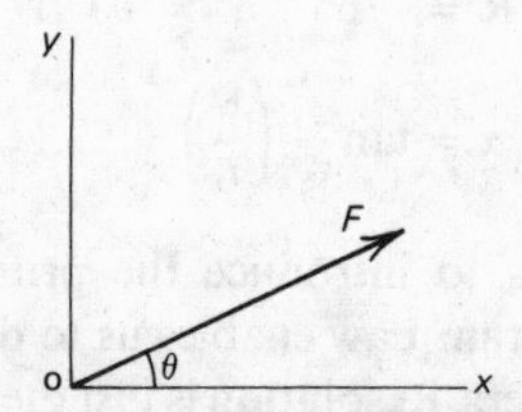

10

The required components are

$$F_x = F \cos \theta$$
$$F_y = F \sin \theta$$

All this work on single forces and resultants doesn't seem to have much to do with Statics. But now look back to Frame 5 where we found the resultant of forces of 2 kN and 3 kN.

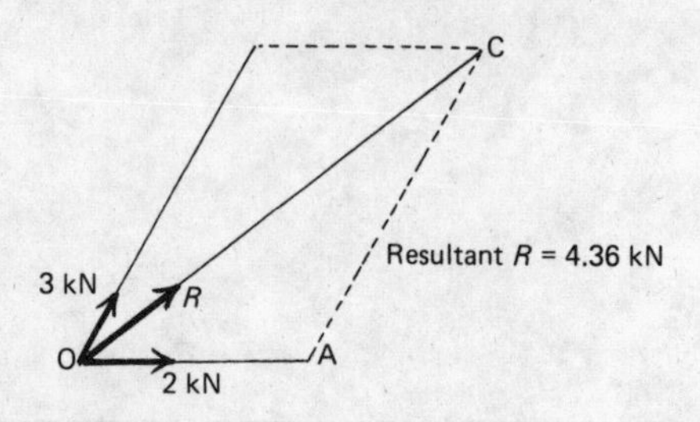

Now if we add to the 2 kN and 3 kN forces a third force of 4.36 kN acting in the *opposite direction* to R, this would cancel the two forces, and the system of three forces would then be in equilibrium. The additional force required in order to produce equilibrium of a force system is called the *equilibrant*, and it is, of course, equal in magnitude to the resultant, and opposite in direction. Thus, the force system

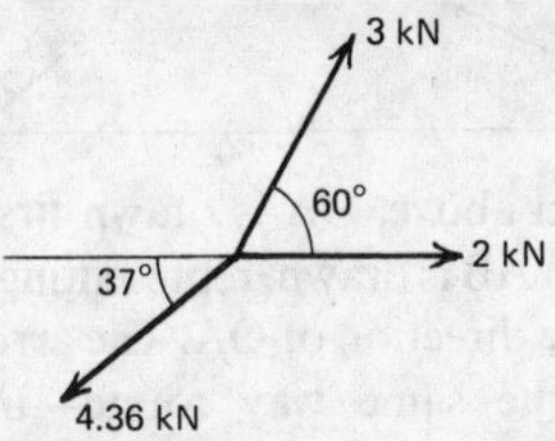

must be in equilibrium.

But these three forces can still be represented by a triangle which is geometrically identical to triangle OAC of Frame 5, reproduced above

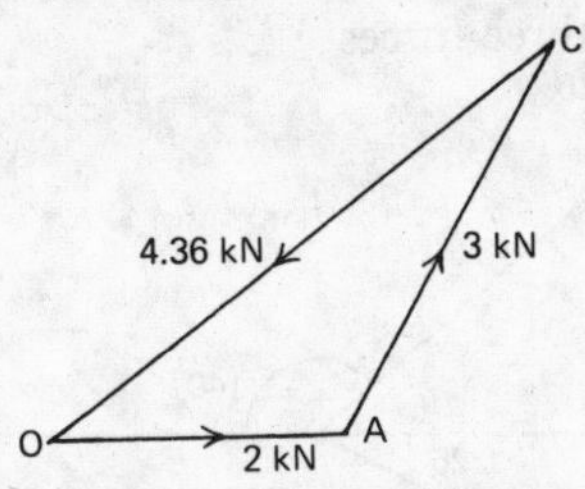

The difference is, that the line CO (read in that order, not OC) now represents a third force, acting in that direction, and not a resultant force, as OC did before. The arrow on CO indicating the direction of the force is seen to follow the same direction as the other two arrows around the triangle. This leads us to the statement of the next fundamental principle of Statics—the Triangle of Forces.

> "Three co-planar forces in equilibrium acting at a point may be represented in magnitude and direction by the three sides of a triangle"

Use this principle now to find graphically the magnitude and direction of a third force to be added to these two

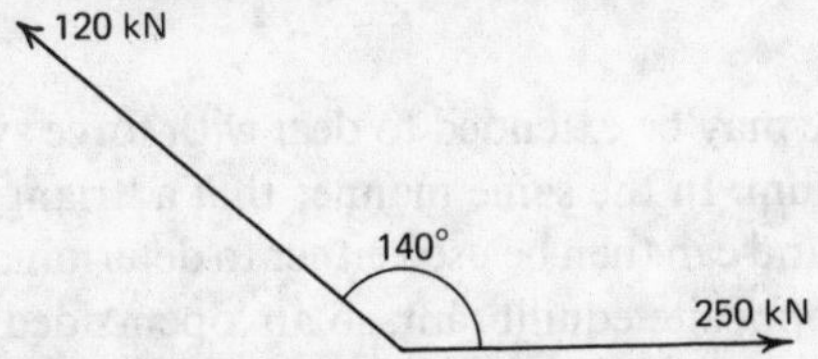

to establish equilibrium. Remember that your completed triangle must have the arrows all running the same way round the triangle.

11

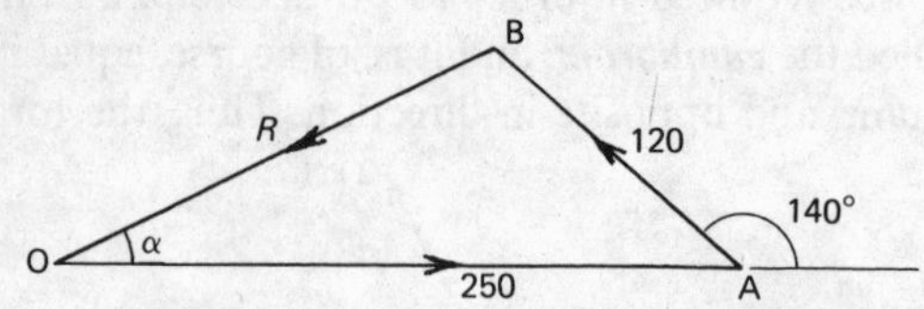

The solution is drawn above. OA is drawn first, 250 units long, to a convenient scale. Then AB is drawn, representing the force of 120 kN. It is drawn at 140° to the direction of OA. The arrows on these two force 'vectors' must run the same way round—in this example, anti-clockwise—round the triangle. BO completes the triangle. Scale measurement gives a value of 176 units, and its direction (with respect to OA) is given by angle α which measures approximately 26°. So the following system of three forces

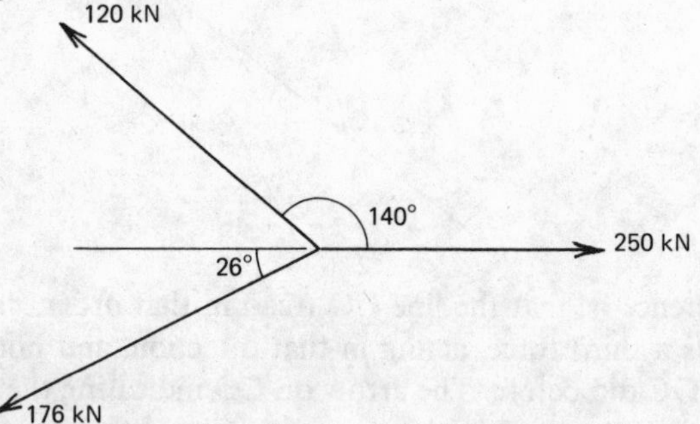

will be in equilibrium.

(If your solution agreed with these figures to within 3 or 4 per cent of error, this may be considered sufficiently accurate for a graphical exercise.)

The principle of the Triangle of Forces may now be extended to the more general Polygon of Forces. Thus

> "A number of co-planar forces acting at a point, if in equilibrium, can be represented, in magnitude and direction, by the sides of a closed polygon"

The principle may be extended to deal with force systems which are not in equilibrium. In the same manner that a 'triangle' can be drawn for two forces, and can then be used either to determine the resultant, or the equal and opposite equilibrant, so an 'open-sided' polygon can be drawn for any number of co-planar forces. To illustrate this, consider three forces acting at a point.

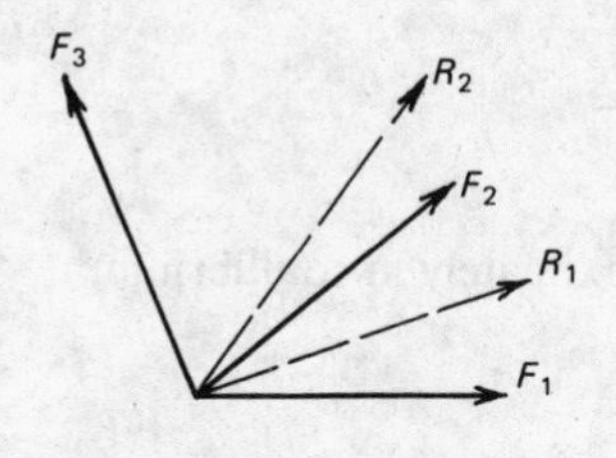

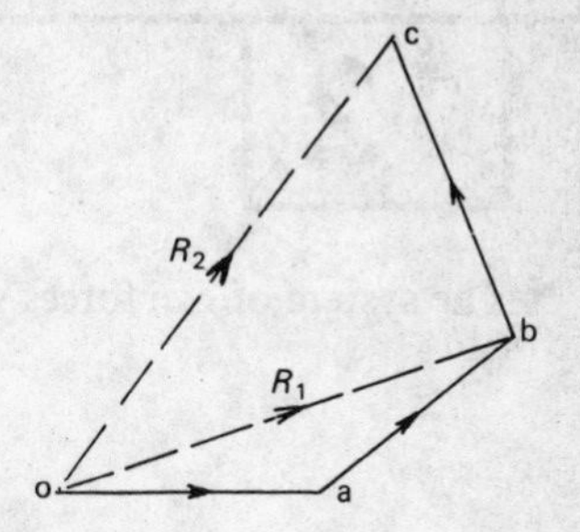

The resultant of forces F_1, and F_2 can be found by drawing the 'open' triangle oab, giving the force R_1. So F_1 and F_2 can be replaced by the single force R_1. The resultant of R_1 and F_3 is found by drawing the 'open' triangle obc, giving R_2 which is the resultant, therefore, of all three forces. It is not necessary to determine R_1; the 'open' polygon oabc can be drawn straight away to determine oc, the resultant vector of the three forces, or co, the equilibrant. Clearly the argument may be extended to include any number of forces.

12

The system of five forces shown here is approximately in equilibrium.

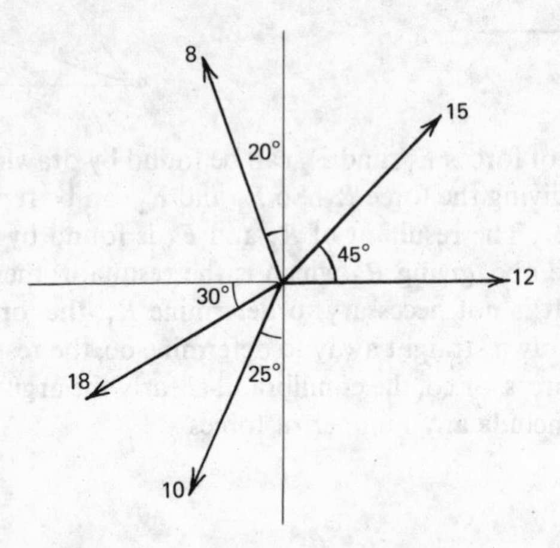

Verify this graphically by drawing a polygon of forces. Choose a suitable scale, remembering that too small a scale means loss of accuracy. The smallest scale you should choose should be 1 cm ≡ 4 units. It should be realised that you may draw the polygon taking the forces in *any order*—which incidentally means that in this exercise there are 24 possible polygons. However, the solution given in the next frame starts with the 12-unit force and takes them in order anti-clockwise around the point, so for a start you might do well to do the same. You may also find it helpful, when drawing the sides of the polygon, to calculate the angle between one force and the next. For example, the angle between the 12-unit force and the 15-unit force is 45° (as shown); the angle between the 15-unit force and the next 8-unit force is 65°, and so on.

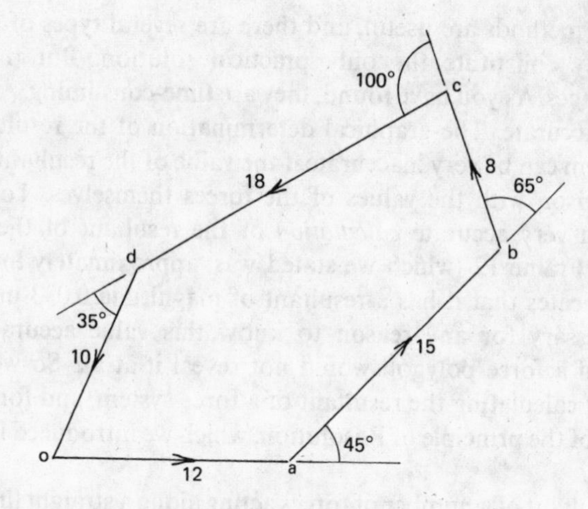

If you choose to draw in a different order, here are sketches of four of the 24 possible polygons.

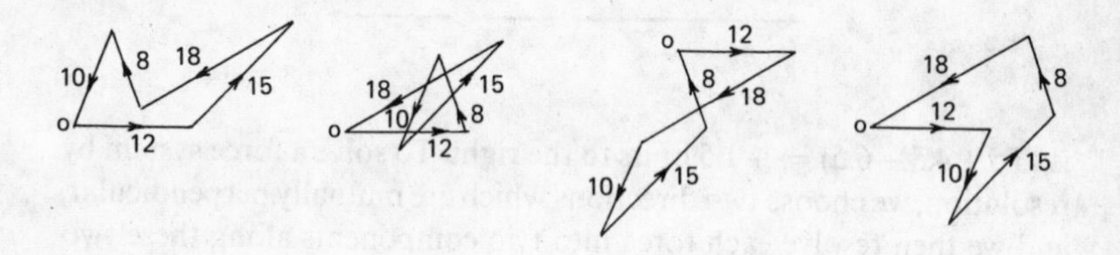

14

Graphical methods are useful, and there are several types of problem where they constitute the only practical solution. But they have disadvantages. As you have found, they are time-consuming. Also, they can be inaccurate. The graphical determination of the resultant of a force system can be very inaccurate if the value of the resultant is small in comparison with the values of the forces themselves. To give an example: a very accurate *calculation* of the resultant of the 5-force system of Frame 12 (which we stated was 'approximately in equilibrium') indicates that it has a resultant of magnitude 0.083 units. If it were necessary for any reason to know this value accurately, the drawing of a force polygon would not reveal it at all. So we need a method of calculating the resultant of a force system, and for this, we make use of the principle of Resolution, which we introduced in Frame 7.

The resultant of a number of forces acting along a straight line is easy to calculate; we simply evaluate the algebraic sum of the forces in one direction. Thus, the resultant of the following three forces

is $(3.5 + 4.5 - 6.5) = +1.5$ units to the right. To solve a force system by resolution, we choose two directions which are mutually perpendicular, and we then resolve each force into two components along these two directions. The algebraic sum of all the components in each of the two directions is then determined; this gives the two components of the required resultant force. From these, the magnitude and direction of the resultant is calculated. The following simple example illustrates the method. We require the resultant of the two forces

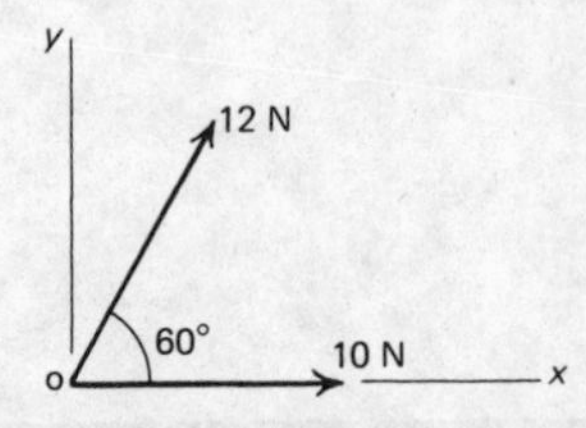

We have chosen the directions ox and oy. There is no need to resolve the 10 N force; it already lies along the ox direction. Making use of the principles of Frames 7 to 9, calculate the two components of the 12 N force.

15

For the 12 N force

$$F_x = 12 \cos 60° = 6.00 \text{ N}$$
$$F_y = 12 \sin 60° = 10.39 \text{ N}$$

So the system now comprises

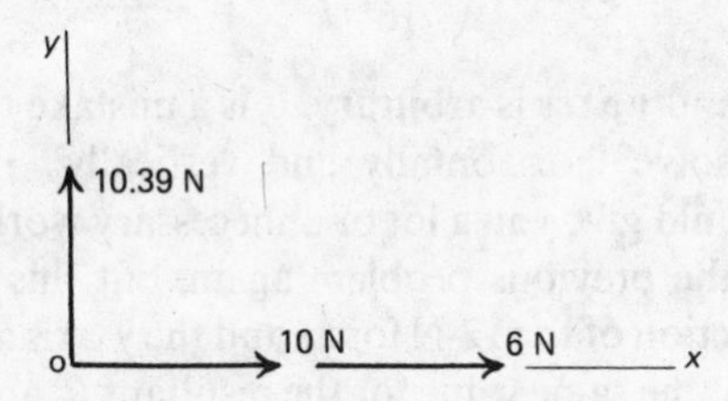

which can be simplified, on adding the x-forces, to

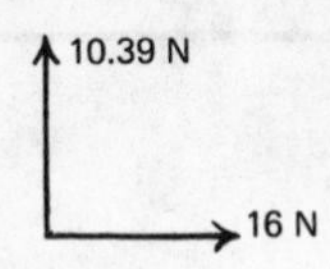

Find the magnitude and direction of the resultant of these two components. Look back to Frame 7 if you are stuck.

16

Sketching the 'triangle' (which is right-angled)

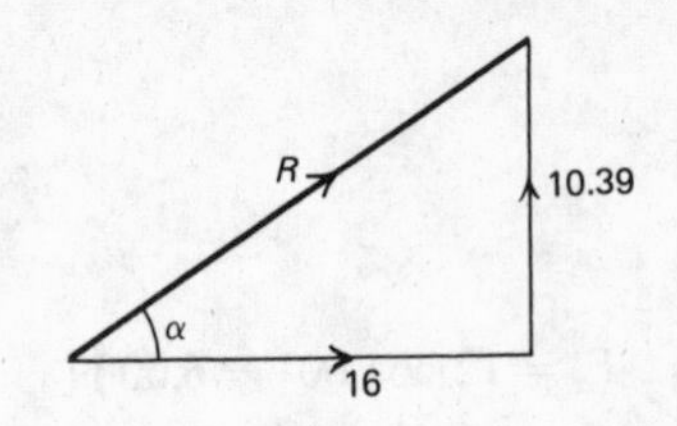

$$R = \sqrt{(16)^2 + (10.39)^2} = \underline{19.08\ \text{N}}$$

$$\alpha = \tan^{-1}\left(\frac{10.39}{16}\right) = \underline{33.00^\circ}$$

The choice of x and y axes is arbitrary. It is a mistake to think that you must always resolve 'horizontally and vertically'. In solving some problems, this could give you a lot of unnecessary work. Let us now go back and solve the previous problem again, but this time taking the x-axis in the direction of the 12 N force, and the y-axis at right-angles to it. You should get the same value for the resultant R, and it should be in the same direction. The solution follows, in case you run into difficulty.

17

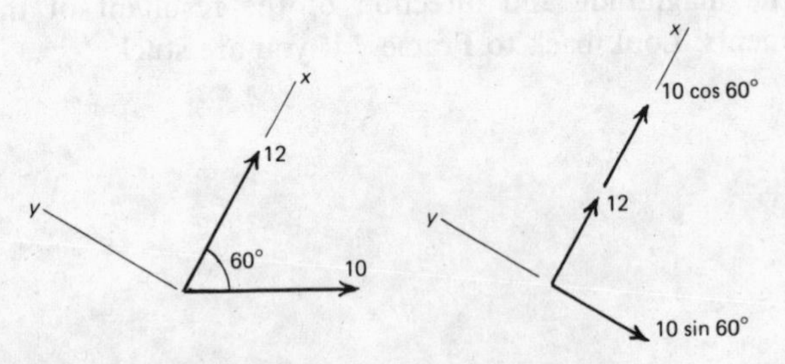

Resultant force in x-direction: $F_x = 12 + 10 \cos 60^\circ = 17.00$ N
Resultant force in y-direction: $F_y = -10 \sin 60^\circ = -8.66$ N
Sketching the triangle

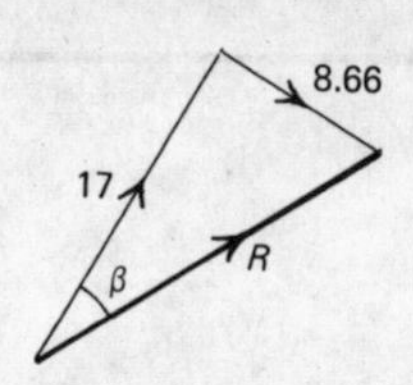

$$R = \sqrt{(17)^2 + (8.66)^2} = \underline{19.08 \text{ N}}$$

$$\beta = \tan^{-1}\left(\frac{8.66}{17}\right) = \underline{27.00^\circ}$$

So the inclination of the resultant R to the horizontal is $(60 - 27) = 33^\circ$ which is the same answer as previously.

Here is another problem. Find the resultant of the four forces shown. The choice of x and y axes should be clear in this case. The solution is given in the next frame, and Frame 19 follows this with some 'drill' examples.

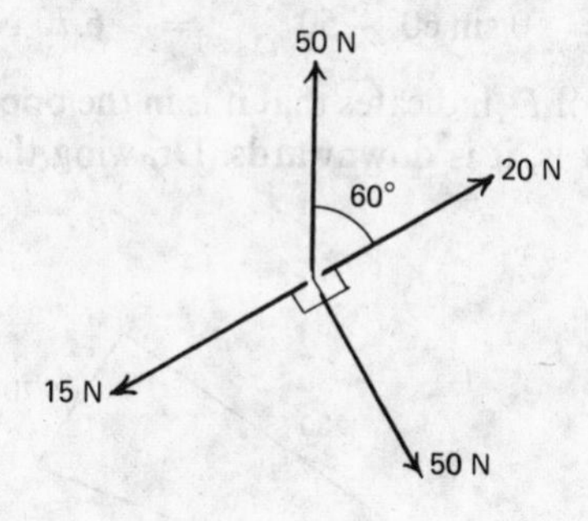

18

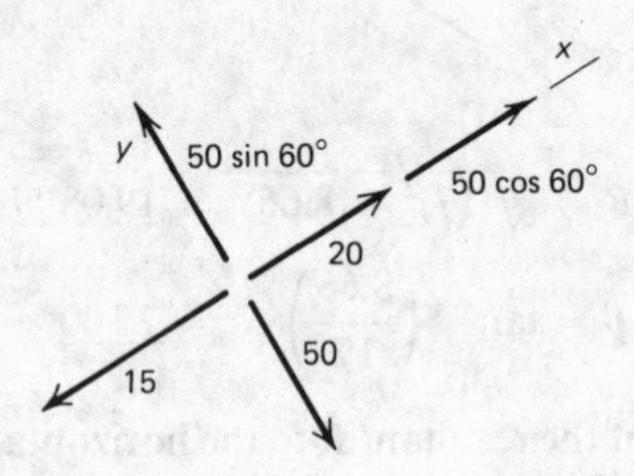

This choice of x and y axes means that only the vertical force of 50 N needs to be resolved.

$$F_x = 20 + 50\cos 60° - 15 = +30\text{ N}$$
$$F_y = 50\sin 60° - 50 = -6.70\text{ N}$$

The negative value of F_y indicates that it is in the opposite direction to that assumed—that is, it is downwards. Drawing the final triangle

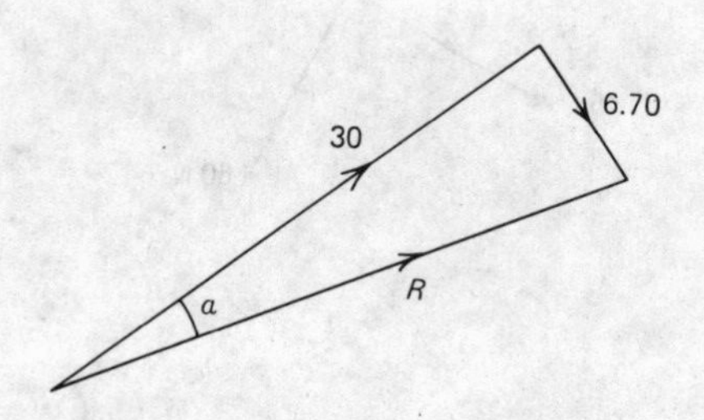

$$R = \sqrt{(30)^2 + (6.70)^2} = \underline{30.74\text{ N}}$$
$$\alpha = \tan^{-1}\left(\frac{6.70}{30}\right) = \underline{12.59°}$$

The resultant force is therefore inclined upwards to the horizontal at an angle of $(30 - 12.59) = 17.41°$.

You may have solved this problem differently, possibly by choosing a different set of x and y axes; for example, you may have chosen horizontal and vertical axes. Also, you may have reached the correct answers. But if you have done so, you will almost certainly find that your working is more complicated than the solution given here. It is a good thing to learn (if you haven't learned already) that there is no single 'right' way to solve a problem, but there are easy ways and less easy ways.

19

'Drill' exercises: forces acting at a point

1. Determine the magnitude and direction of the resultant of two forces acting at the same point, of magnitudes 64 N and 96 N, angularly displaced relatively by 60°.
   Ans. 140 N at 23.5° to 96 N force.
2. Forces of 200 N and 160 N act at a point at a relative angle of 120°. Calculate the magnitude and direction of a third force required to produce equilibrium with these forces.
   Ans. 183 N at 130.8° to 200 N force.
3. Forces of 4, 5 and 4.5 N act at a point in relative directions of 0°, 45° and 135°. Determine the magnitude and direction of their resultant.
   Ans. 8.01 N at 12° to 5 N force.
4. Forces of 8, 5, 6 and 6 kN act at relative angles of 0°, 90°, 120° and 210°. Determine the magnitude and direction of the resultant.
   Ans. 7.2 kN at 1.6° to 5 kN force.
5. A street lamp weighing 100 N hangs from an initially horizontal wire stretched between posts 10 m apart. The weight of the lamp causes the wire to sag 0.2 m at the centre. Calculate the tension in the wire.
   Ans. 1251 N.
6. A body of mass 20 kg hangs from two strings which are inclined to the horizontal at angles of 30° and 45° respectively. Calculate the tensions in the strings.
   Ans. 176 N and 143.6 N.
7. Three forces, A, B and C, act at a point at respective angles of 0°, 135° and 270°. The magnitude of A is 200 N. Given that the three forces are in equilibrium, calculate the values of B and C.
   Ans. 282.8 N and 200 N.
8. A body of mass 5 kg hangs from two cords of lengths $3\frac{1}{2}$ m and $4\frac{1}{2}$ m. The other ends of the cords are attached to a vertical board to points at the same level and $5\frac{1}{2}$ m apart. Calculate the tensions in the two strings.
   Hint: a graphical solution is the simplest.
   Ans. 38.0 N and 28.4 N.
9. ABCD is a rectangle. AB is 3 m, BC is 4 m. E is a point on BC $2\frac{1}{4}$ m from B. Forces of 6, 8, 10 and 12 kN act along AB, AE, AC and AD

respectively. Determine the magnitude and direction of their resultant.

Ans. 30.9 kN at 53.4° to AB.

10. Bodies of weight, 3 kN and 4 kN are connected by a light cord which passes over pulleys on a vertical board as shown. A third body of 6 kN is attached to the cord. Calculate the angles α and β when the bodies take up a position of equilibrium. Neglect friction.

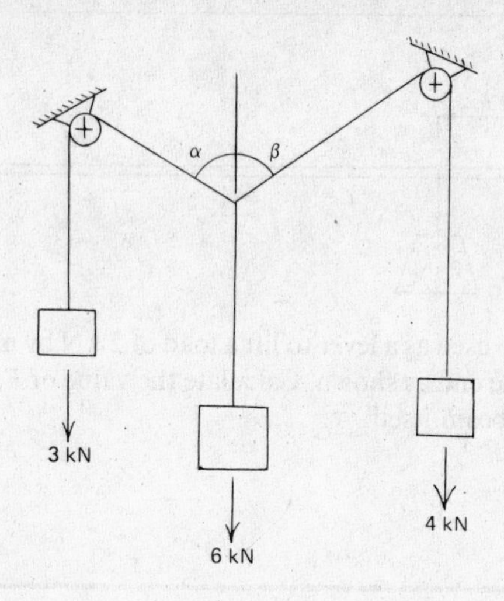

Hint: draw the force triangle for the three forces at the junction of the cords. Solve either graphically or by calculation, using the sine and cosine formulae.

Ans. 36.3° and 26.4°.

20

Now have a look at this simple force problem.

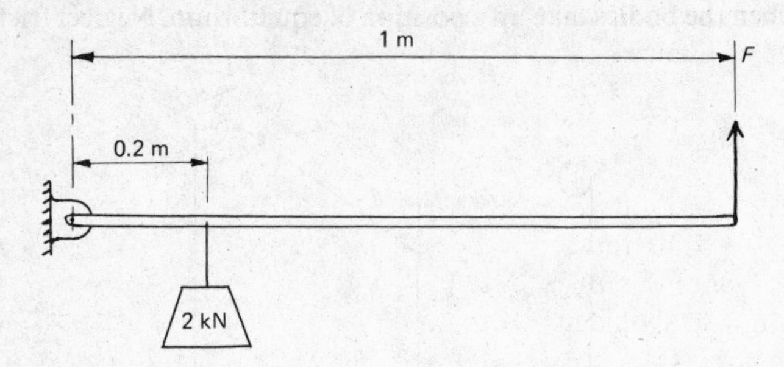

A beam is to be used as a lever to lift a load of 2 kN by means of a force F applied at the end as shown. Calculate the value of F, neglecting the weight of the beam itself.

21

You may know that this problem can be solved by taking moments of forces about the hinge-point at the left-hand end. Although this is a simple problem, it is different from any that we have so far examined, because the forces concerned do not all pass through a single point, with the result that the methods we have so far used cannot be employed to solve this problem.

When writing an equation of moments, we must note that the force F, if acting alone, would cause the beam to turn anti-clockwise, whereas the weight of the load would cause it to turn the other way. Moments, like forces, must carry positive or negative signs, according to direction. Since the beam is in a state of equilibrium, the total moment is zero. Arbitrarily calling an anti-clockwise moment positive

$$F \times 1 - 2 \times 0.2 = 0$$

giving

$$F = 0.4 \text{ kN}$$

We must think of a moment of a force as its turning effect with respect to a particular point. The magnitude of a moment is calculated by multiplying the value of the force by the perpendicular distance from the point to the *line of action* of the force. If the force F were applied to the beam at a distance of only 0.5 m from the hinge, the value of the moment would thereby be halved, and to ensure equilibrium, the magnitude of F would thus have to be doubled in order to raise the same load. Similarly, if the weight were to be moved out to a distance of 0.4 m from the hinge, the moment of the weight would be doubled, and again, F would have to be doubled to compensate.

You must pay particular attention to the method of calculating moment, as set out in the previous paragraph. A very common error associated with this type of problem is to consider the point of application of the force rather than its line of action. But if you think about it, you should see the fallacy of this. Imagine the force F to be applied by tying a rope to the beam and pulling the end of the rope. Regardless of whether the length of the rope was 0.1 m, 1 m or 10 m, you would need to apply a force of 0.4 kN to raise the load. What is important is that the rope be vertical, not how long it was. Although the point of application of force F (the end of the rope) changes, the line of action of F (a vertical line through the end of the beam) remains unaltered, and therefore the value of F remains the same.

Bearing this very important principle in mind, see if you can now calculate the value of the force F required to raise the weight in the example below.

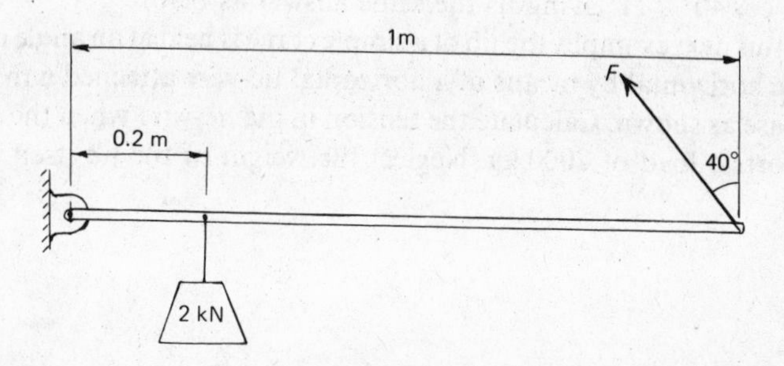

22

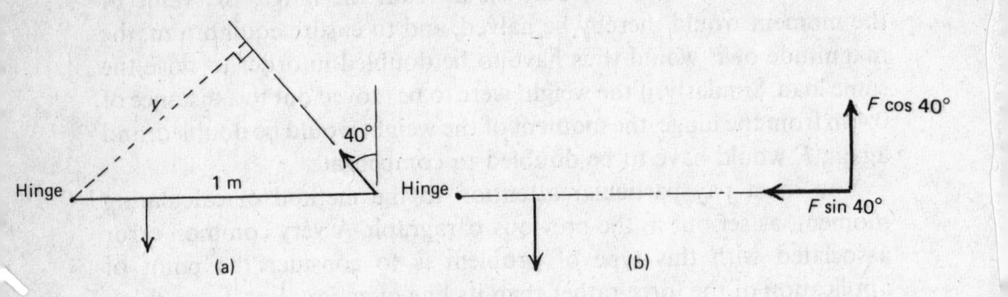

In diagram (a) the required perpendicular distance between the line of action of F and the hinge, shown as y, is $1 \cos 40°$, so that we can write the moment equation

$$F \times 1 \cos 40 = 2 \times 0.2$$

$$\therefore\ F = \underline{0.522\ \text{kN}}$$

Diagram (b) suggests an alternative solution. Here, the force F has been *resolved* into horizontal and vertical components (see Frame 8). The line of action of the horizontal component actually passes through the hinge, and so has a moment of zero about it. The vertical component, $F \cos 40°$, being perpendicular to the beam, has a moment of $(F \cos 40° \times 1)$, giving us the same answer as before.

In this next example, the jib of a simple crane is held at an angle of 40° to the horizontal by means of a horizontal tie-wire attached 6 m from the base as shown. Calculate the tension in the tie-wire when the crane supports a load of 2000 kg. Neglect the weight of the jib itself.

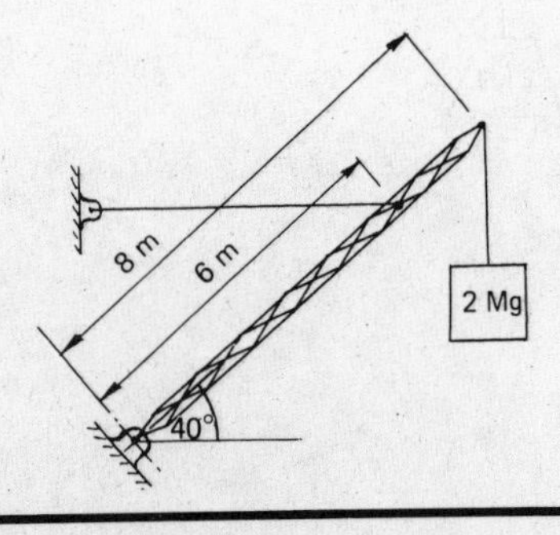

23

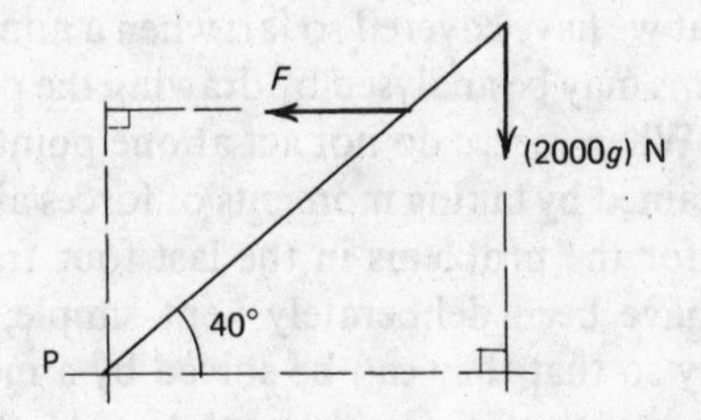

Calling the tensile force in the wire F, we write an equilibrium equation of moments of the two forces with respect to the point P, the foot of the jib, recalling that the weight of a body, in newtons, is its mass, in kilograms multiplied by g (9.81)

$$F \times 6 \sin 40^\circ - 2000g \times 8 \cos 40^\circ = 0$$

$$\therefore\ F = \left(2000g \times \frac{8}{6} \cot 40^\circ\right) \mathrm{N}$$

$$= \underline{31.18 \text{ kN}}$$

You should take note of the fact that the force diagram above is simplified, because the forces acting on the jib also include a force exerted upon it by the mounting at the foot P. But since we are writing a moment equilibrium equation with respect to P, the moment of this third force above P must be zero, and we can thus solve the problem without needing to know this force.

24

Summarising what we have covered so far: when a number of forces act at a point, problems may be analysed by drawing the polygon of forces, or by resolution. When forces do not act at one point, a solution may sometimes be obtained by taking moments of forces about some point. (This is the case for the problems in the last four frames.) But these latter problems have been deliberately kept simple; they have been chosen specifically so that they can be solved by a moment equation. But suppose now we want a *complete* solution to this later kind of problem. For example, suppose that in addition to the tension in the stay-wire of the jib of the last frame, we want to know the magnitude and direction of the forces exerted on the jib by the anchor-bracket at the base P. Then we need to make use of the *two* conditions required for equilibrium, which are stated here.

1. There shall be no resultant force on the body.
2. The sum of moments of all forces acting on the body, with respect to any point must be zero.

We shall now look at a type of problem in which all the forces are parallel. Here is one.

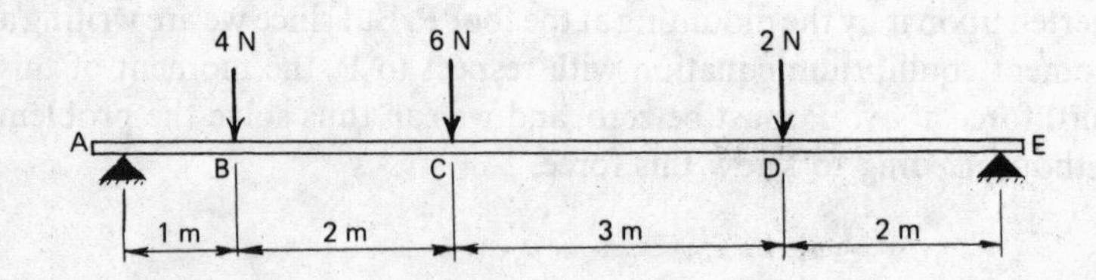

A light bar ABCDE carries three loads as shown, and rests on two knife-edge supports at the ends A and E. Determine the values of the upward reaction forces at A and E.

We can call the two reaction forces R_A and R_E. Our force system becomes

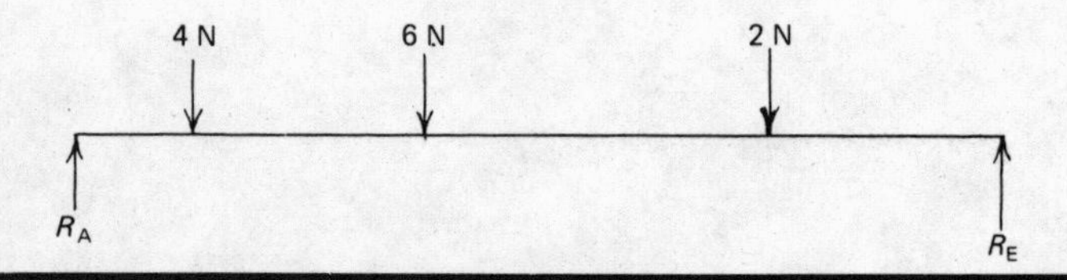

Since all the forces are parallel, we do not need to resolve any forces, and there is even less point in drawing a force polygon—which, in any case, would be just a single straight line. Our first condition of equilibrium can be stated in terms of an equation of vertical equilibrium, namely

$$R_A + R_E = 4 + 6 + 2 = 12\ \text{N}$$

Because the forces are not all acting at a single point, this equation by itself is insufficient to solve the problem. A second equation, making use of the Moment principle is required. You may obtain such an equation by taking moments of the forces about any point. But the solution is made simpler by using some discretion. By choosing your point carefully, the working is made much simpler. Think about this and then write down a Moment equation. The working continues in the next frame.

25

Observe the result of taking moments about the point E.

$$R_A \times 8 - 4 \times 7 - 6 \times 5 - 2 \times 2 = 0$$

(Note here, as previously, that moment can be positive or negative, and in this equation, we have arbitrarily chosen to represent a moment with a clockwise sense as positive, and vice versa.)

You can see that by choosing point E, the force R_E does not come into the equation: the moment of R_E about E is $R_E \times 0 = 0$. The equation may thus be solved.

$$R_A = \frac{1}{8}(4 \times 7 + 6 \times 5 + 2 \times 2) = \underline{7\tfrac{3}{4}\ \text{N}}$$

A second equation could now be written, taking moments this time about point A, and from this we could find the value of R_E. But we do not need to do this, as we have the equation at the end of Frame 24, from which

$$R_E = 12 - R_A = 12 - 7\tfrac{3}{4} = \underline{4\tfrac{1}{4}\ \text{N}}$$

The next problem is slightly more difficult, but exactly the same principles are required. Find the values of the reaction forces at points B and F.

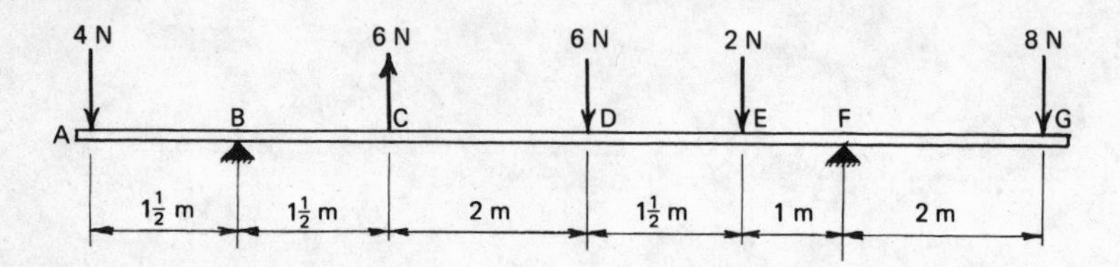

Begin by taking moments about point F. Call clockwise moments positive. Then write the equation of vertical equilibrium (note that the load at C is upwards).

26

Taking moments of all forces about F

$$-4 \times 7\tfrac{1}{2} + R_B \times 6 + 6 \times 4\tfrac{1}{2} - 6 \times 2\tfrac{1}{2} - 2 \times 1 + 8 \times 2 = 0$$

from which

$$R_B = \tfrac{1}{6}(4 \times 7\tfrac{1}{2} - 6 \times 4\tfrac{1}{2} + 6 \times 2\tfrac{1}{2} + 2 \times 1 - 8 \times 2) = \underline{\tfrac{2}{3}\text{ N}}$$

Writing the equation of vertical equilibrium

$$R_B + R_F = 4 - 6 + 6 + 2 + 8 = 14$$
$$\therefore\ R_F = 14 - \tfrac{2}{3} = \underline{13\tfrac{1}{3}\text{ N}}$$

For our next exercise, go back to the three loads on the bar in Frame 24. But instead of the bar being supported at A and E, imagine it to be balanced at a single point. The problem is to locate this point—the point at which the bar would balance on a single knife-edge. We may call the point F, and assume it to be at a distance X from A. Our force system is thus

Begin with a moment equation about A.

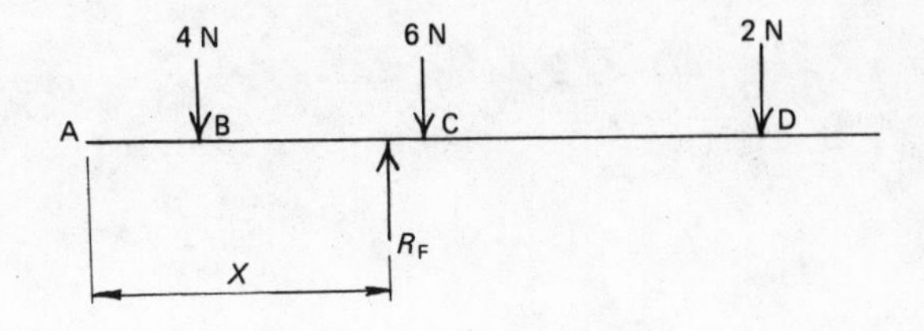

27

$$R_F \times X = 4 \times 1 + 6 \times 3 + 2 \times 6 = 34$$

This cannot be solved as it stands, but a second equation of vertical equilibrium can be written.

$$R_F = 4 + 6 + 2 = 12$$

$$\therefore X = \frac{34}{12} = \underline{2\tfrac{5}{6}\ \text{m}}$$

In the earlier problems, the solution was made simpler by a careful choice of which point to take moments about. But in this case it is not important. You can verify the solution above by solving the problem again, but this time assuming the point F to be a distance Y from B, and then taking moments about B. You can thus show that $Y = \underline{1\tfrac{5}{6}\ \text{m}}$ which locates F in the same position as before.

28

The last several examples have all been cases of force systems in equilibrium. We can now extend the principle of moments to include the determination of the *resultant* of a number of parallel forces which are not in equilibrium. To do this, we need to modify the two statements made in Frame 24. In finding the resultant of a system of parallel forces, the following conditions apply.

1. The magnitude of the resultant is equal to the algebraic sum of the forces.
2. The moment of the resultant with respect to some point is equal to the algebraic sum of the moments of all the forces with respect to the same point.

The word 'algebraic' in these statements reminds you that both force and moment have signs. If an upward-acting force is designated positive, then a downward-acting one must be negative. Similarly, if force directed to the left is positive, then force to the right must be negative. And, as we have seen, if we call a clockwise moment positive we must take care to call an anti-clockwise moment negative.

The first of the two statements above enables you to calculate the *magnitude* of the resultant; the second enables you to determine its *line of action.*

Let's look again at the example in Frame 24, and determine the resultant of the three downward-acting loads on the beam. (You should see that we cannot find the resultant of *all* the forces acting on the beam, because this will be zero, the beam being in equilibrium.) We shall call the resultant R, and shall assume that its line of action is a distance X from point A. On the diagram below, we show the force between C and D, but of course we do not know this; this is just a guess

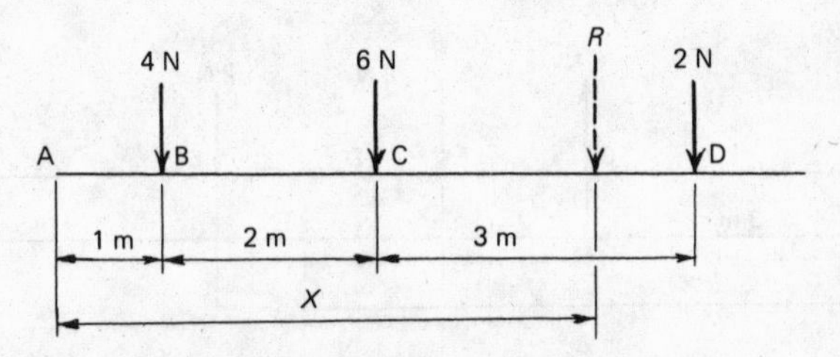

Solve the problem by taking moments about point A.

29

The moment equation about A is

$$R \times X = 4 \times 1 + 6 \times 3 + 2 \times 6 = 34$$

and, from the first condition stated in Frame 28

$$R = 4 + 6 + 2 = 12$$

from which

$$X = \frac{34}{12} = 2\tfrac{5}{6}\ \text{m}$$

This answer may be familiar; it is the same as that obtained to the problem in Frame 27. But in that case, we were finding the single upward force required on the loaded beam to produce equilibrium, whereas here, we are finding the resultant of the three loads. As we have seen in several examples so far, the resultant of any force system will always be equal in magnitude to the force required to produce equilibrium, will be opposite in direction, and will lie along the same straight line.

Using exactly the same technique, now see if you can determine the resultant of the five loads on the beam in Frame 25. Assume that its line of action is a distance X from the end A of the beam. The full solution follows in Frame 30.

30

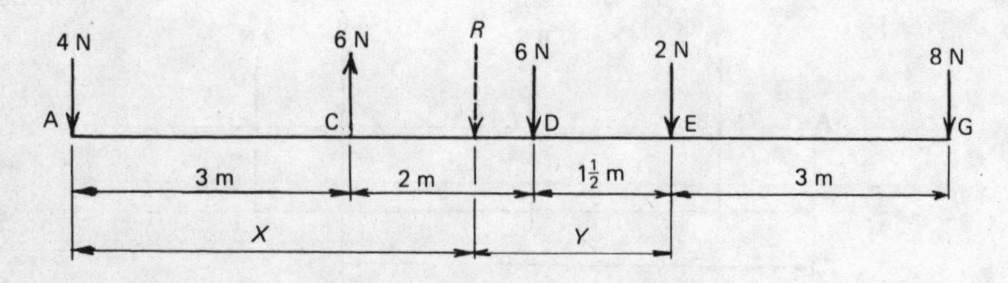

The equation of moments about A is

$$R \times X = 4 \times 0 - 6 \times 3 + 6 \times 5 + 2 \times 6\tfrac{1}{2} + 8 \times 9\tfrac{1}{2} = 101\ \text{N}$$

Equating R to the algebraic sum of forces

$$R = 4 - 6 + 6 + 2 + 8 = 14$$

giving

$$X = \frac{101}{14} = 7\tfrac{3}{14}\text{ m}$$

which locates the resultant between E and G, $\frac{5}{7}$ m to the right of E, and not between C and D as guessed in the diagram.

Again, you can verify this solution by solving the problem again, taking moments about another point. If, for example you take moments about point E, you should be able to show that the line of action of R is again $\frac{5}{7}$ m to the right of E. If you use the same diagram as the one above, calling the required distance Y, as shown, you would get a negative value for Y. You can give yourself as much practice as you need by solving the problem by taking moments about any other point you wish.

31

It is a mistake to think that parallel force systems must always consist of vertical forces only. Find the resultant of these three forces. Assume the line of action of the resultant, R, to be a distance Y from point A as shown.

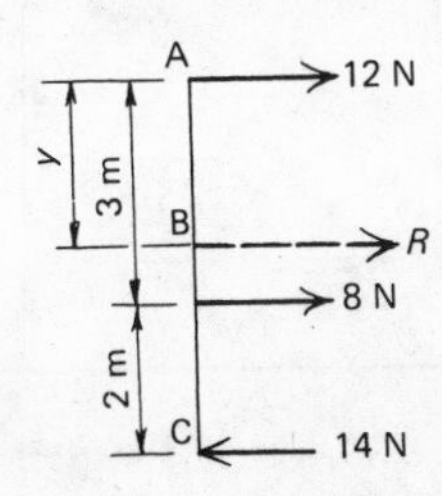

32

Taking moments about A

$$R \times Y = 12 \times 0 + 8 \times 3 - 14 \times 5 = -46$$

Equating R to algebraic sum

$$R = 12 + 8 - 14 = +6\text{ N}$$

Thus

$$Y = \frac{-46}{6} = \underline{-7\tfrac{2}{3}\text{ m}}$$

The negative sign now tells us that the line of action of R is *above* A—perhaps a rather surprising result.

Sometimes we may have a combination of horizontal and vertical forces. In the problem that follows, we have eight forces acting at the corners of a rectangle ABCD which is 4 m by 6 m. We require the resultant of all eight forces.

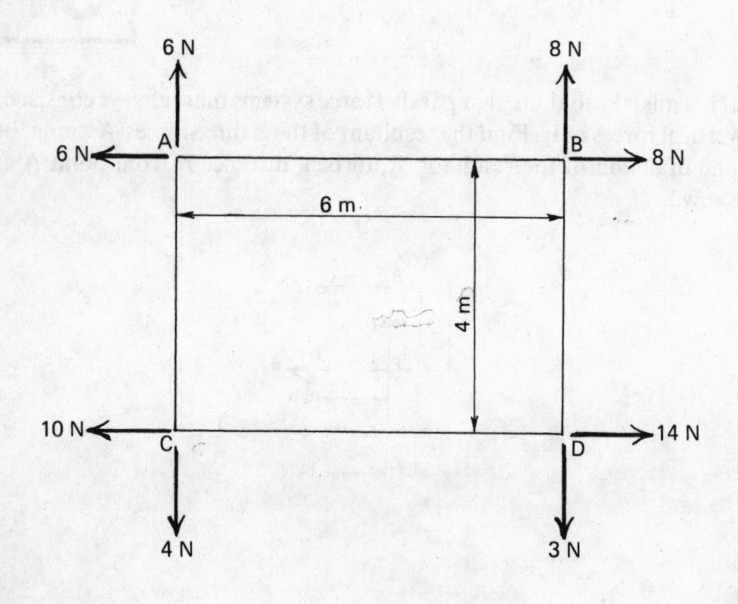

We solve this by determining the resultants of the system of horizontal forces, and of the vertical force system, separately. Call these respectively R_H and R_V. Calculate the values of these, and find their lines of action by assuming the line of action of R_H to be a distance X above D, and that of R_V to be a distance Y to the right of A. You should find that $X = 1\frac{1}{3}$ m and $Y = 4\frac{2}{7}$ m but the complete working is given in Frame 33.

33

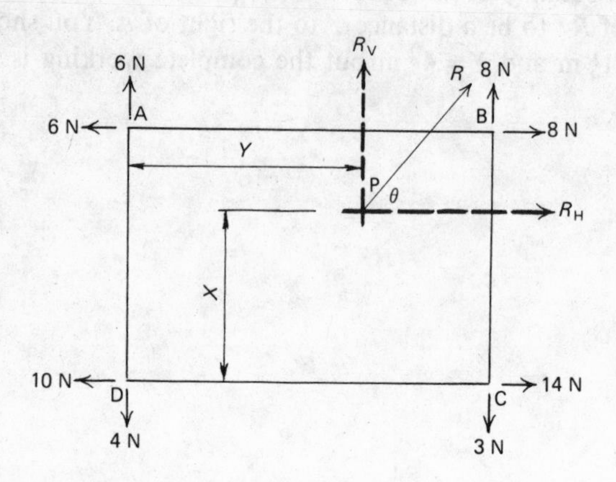

Considering all horizontal forces

$$R_H = 8 - 6 + 14 - 10 = \underline{+6\text{ N}}$$

Now considering all vertical forces

$$R_V = 6 - 4 + 8 - 3 = \underline{+7\text{ N}}$$

The moment equation for all horizontal forces about point C is

$$R_H \times X = 8 \times 4 - 6 \times 4 + 14 \times 0 - 10 \times 0 = \underline{+8\text{ N m}}$$

$$\therefore\ X = \frac{+8}{R_H} = \frac{+8}{+6} = \underline{+1\tfrac{1}{3}\text{ m}}$$

The moment equation for all vertical forces about D is

$$R_V \times Y = 8 \times 6 - 3 \times 6 + 6 \times 0 - 4 \times 0 = \underline{+30\text{ N m}}$$

$$\therefore\ Y = \frac{+30}{R_V} = \frac{+30}{+7} = \underline{+4\tfrac{2}{7}\text{ m}}$$

Now that we know the magnitudes of the two components of the resultant force, we can find the force by compounding the components using Pythagoras, as we did in the problem in Frames 15 and 16.

$$R = \sqrt{R_H^2 + R_V^2} = \sqrt{6^2 + 7^2} = \underline{9.22\text{ N}}$$

and its direction is defined in terms of the angle θ on the above diagram, where

$$\theta = \tan^{-1}\frac{R_V}{R_H} = \tan^{-1}\frac{7}{6} = \underline{49.4°}$$

and this resultant passes through the point P.

This result must be clearly understood. The resultant force of 9.22 N is an imaginary force, and point P is an imaginary point; there is nothing significant about its co-ordinates. A force is defined by three things: its magnitude, its direction, and its line of action. When the first two are known, the line of action can be found by knowing *any one point* through which the line of action must pass. To illustrate this point, let's assume that the resultant force R passes through a point Q situated on CD. Assume as before that it has a direction θ to the horizontal. Begin by resolving R into components R_H and R_V thus

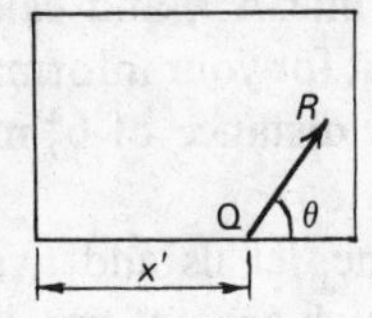

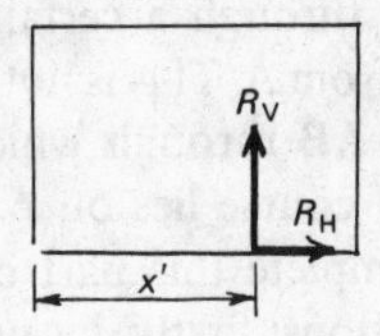

R_H, R_V and R and θ can be determined exactly as before, and of course the values will be as before. We now require the value of distance x'.

34

x' may be calculated from a moment equation about point D

$$R_V \times x' = 6 \times 4 + 8 \times 6 - 8 \times 4 - 3 \times 6 = 22$$

$$\therefore x' = \frac{22}{R_v} = \frac{22}{7} = \underline{3\tfrac{1}{7}\text{ m}}$$

and an examination of the previous solution at the end of Frame 33 and a simple calculation will show that the line of action of R is exactly the same as before. As with so many of these Statics exercises, you can give yourself extra practice by solving the same problem in different ways. You can, for example, solve the problem again but this time assume that R passes through a certain point on the line AB, and calculate its distance from A. This is not done for you, but for your information, the point on AB through which R passes is a distance of $6\tfrac{4}{7}$ m from A (which of course lies on AB produced).

To complete this part of the programme, let us add two further complications; firstly, let us have forces which are not parallel to the sides of the rectangle, and secondly do not let them all be at the corners. Find the magnitude and line of action of the resultant of this system of eight forces.

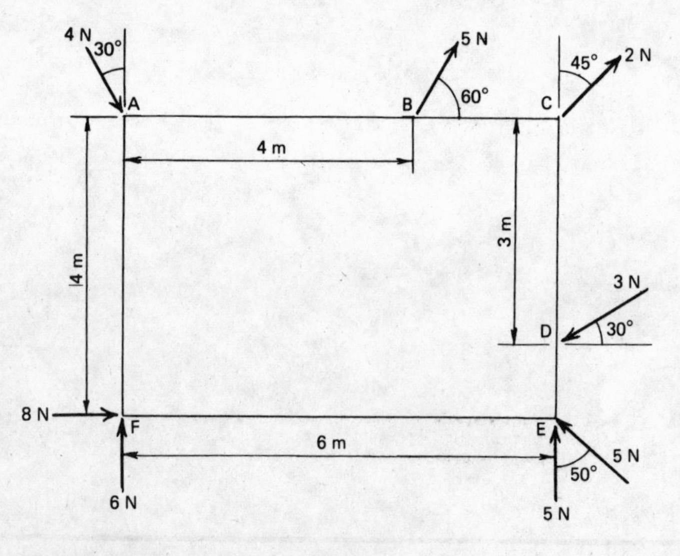

Begin by redrawing the force system, with all inclined forces resolved into components along the horizontal and vertical directions. You can then find the resultant horizontal component, the resultant vertical component, and the magnitude and direction of the resultant. Although the working is shown in Frame 35, the answers to this part of the solution are: $R_H = 8.48$ N to the right; $R_v = 14.99$ N upwards; $R = 17.22$ N at 60.5° to the horizontal.

35

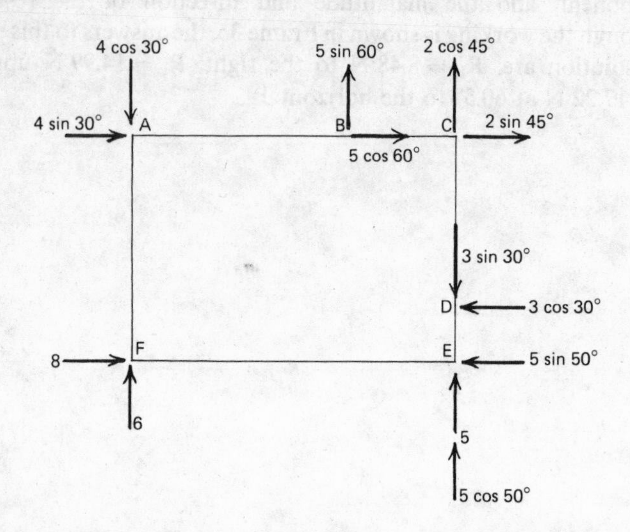

The resultant, R_H, of the six horizontal forces (assumed acting left to right) is

$$R_H = 4\sin 30° + 5\cos 60° + 2\sin 45° - 3\cos 30° + 8 - 5\sin 50°$$
$$= 2.00 + 2.50 + 1.41 - 2.60 + 8.00 - 3.83$$
$$= +8.48\text{ N (that is, to the right, as assumed)}$$

The resultant vertical force R_V (assumed acting upwards) is

$$R_V = -4\cos 30° + 5\sin 60° + 2\cos 45° - 3\sin 30° + 5 + 5\cos 50° + 6$$
$$= -3.46 + 4.33 + 1.41 - 1.50 + 5.00 + 3.21 + 6.00$$
$$= +14.99\text{ N (that is, upwards as assumed)}$$

Assuming R_H to lie along a line at distance Y above EF

$$R_H \times Y = (4\sin 30° + 5\cos 60° + 2\sin 45°) \times 4 - 3\cos 30° \times 1$$
$$= 5.91 \times 4 - 2.6 \times 1$$
$$= 21.04\text{ N m}$$

$$\therefore Y = \frac{21.04}{8.48} = \underline{2.48\text{ m}}$$

Assuming R_V to lie along a line at distance X to the right of AF

$$R_V \times X = 5\sin 60° \times 4 + (2\cos 45° - 3\sin 30° + 5 + 5\cos 50°) \times 6$$
$$= 4.33 \times 4 + 8.12 \times 6$$
$$= 66.04 \text{ N m}$$
$$\therefore X = \frac{66.04}{14.99} = \underline{4.41 \text{ m}}$$

The simplified force system is thus

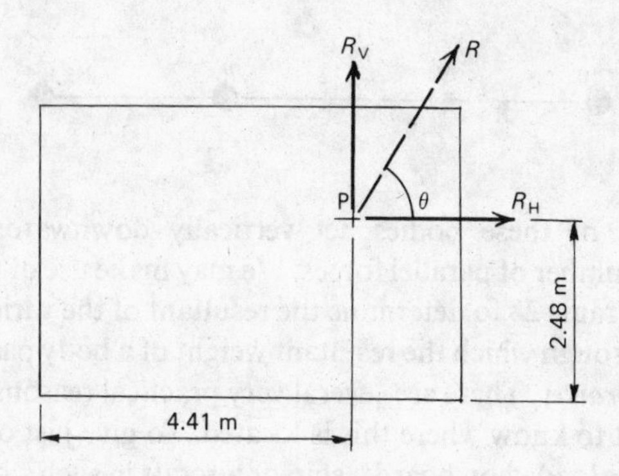

The total resultant, R is given by $R = \sqrt{R_H^2 + R_V^2} = \sqrt{8.48^2 + 14.99^2}$

$$= \underline{17.22 \text{ N}}$$

and angle θ is given by $\theta = \tan^{-1}\left(\frac{14.99}{8.48}\right) = \underline{60.50°}$

36

A particular example of the action of parallel forces is to be found in the consideration of the weight of a body. Let us imagine a system which consists very simply of a number of separate bodies, having various masses, m_1, m_2 etc., concentrated at points, and attached to a light bar of negligible mass. Thus

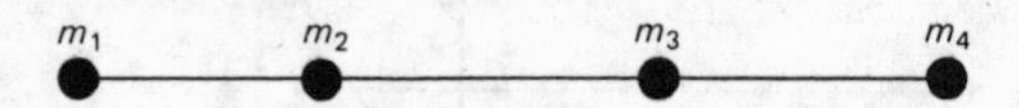

The weights of these bodies act vertically downwards, and thus comprise a number of parallel forces. We may make use of the methods outlined in Frame 28 to determine the resultant of the various weights. The point through which the resultant weight of a body passes is called its *centre of gravity*. There are several very practical reasons why it may be important to know where this is located. To give just one example, cargo must be loaded on board a ship or aircraft in such a manner as to ensure that the centre of gravity of the loaded craft lies on the fore-and-aft axis.

The weight of a body, in newtons, is calculated by multiplying its mass, in kilograms, by 9.81 (g). So the system of parallel forces comprising the weights of the masses on our bar is like this

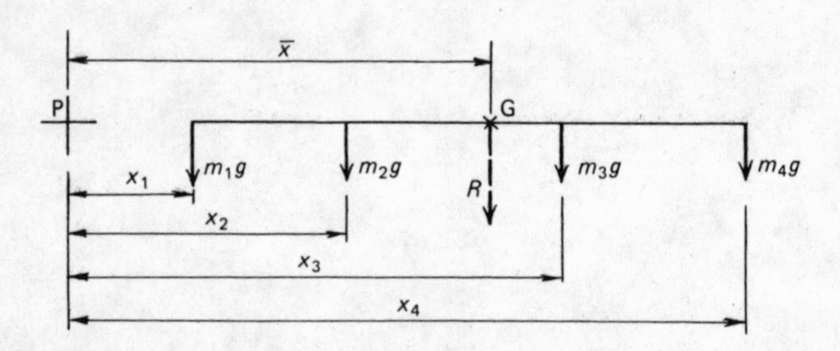

If we assume the resultant weight R to have its line of action a distance $\bar{x}$ to the right of an arbitrary point P, we can apply the Moment principle of Frame 28, thus

$$R \times \bar{x} = m_1 g \times x_1 + m_2 g \times x_2 + m_3 g \times x_3 + m_4 g \times x_4 \text{ etc.}$$

But, from the Force principle of Frame 28, we know that R is the algebraic sum of all the weights. So

$$R = m_1 g + m_2 g + m_3 g + m_4 g \text{ etc.}$$

Thus

$$\bar{x} = \frac{m_1 g x_1 + m_2 g x_2 + m_3 g x_3 + m_4 g x_4 \text{ etc.}}{m_1 g + m_2 g + m_3 g + m_4 g \text{ etc.}}$$

or more concisely

$$\bar{x} = \frac{\Sigma(mx)}{\Sigma(m)} \quad \text{(noting that } g \text{ cancels throughout)}$$

this last expression being in a general form, to allow for *any* number of bodies, and not just four as shown in our example. Have a go at using the formula to find the distance of the centre of gravity of the following system from the left-hand end.

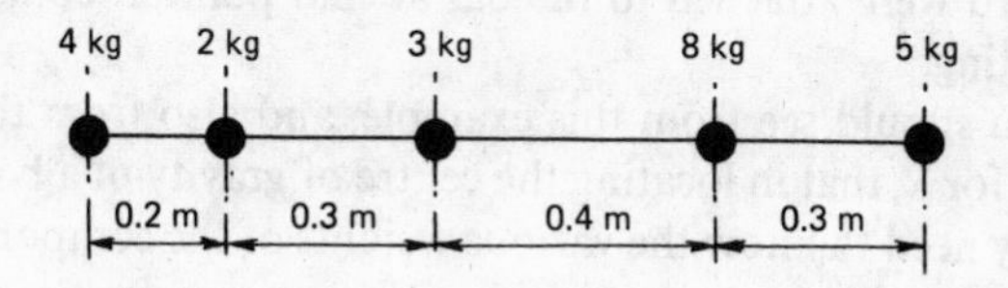

37

Since the problem is to find the distance of the centre of gravity, G, from the left-hand end, we locate our point P at this point, and measure all our distances from there. So the expression for x is

$$\bar{x} = \frac{\Sigma(mx)}{\Sigma(m)} = \frac{4 \times 0 + 2 \times 0.2 + 3 \times 0.5 + 8 \times 0.9 + 5 \times 1.2}{4 + 2 + 3 + 8 + 5}$$

$$= \frac{15.1}{22}$$

$$= \underline{0.686 \text{ m}}$$

This locates the centre of gravity 0.186 m to the right of the 3 kg body, and if a cord were attached to the bar at that point, it could be lifted without tilting.

Now you should see from this example, and also from the general expression for $\bar{x}$, that in locating the centre of gravity of a body, we do not actually need to know the various weights of the component parts. You recall that in the derivation in the previous frame, g actually cancelled out. The formula for $\bar{x}$ is thus seen to be not merely a formula for locating a resultant weight, but also an indication of the manner in which the mass of a system is distributed. (We should get the same answer to the above problem if the system of masses were on the moon, where all the weights would be different from the earth-values—or even if the bar were located in a weight-free field.) For this reason, we rarely use the term 'centre of gravity' but use instead the expression '*centre of mass*' or, more simply, just '*mass centre*'.

38

The previous problem was idealised considerably; we conveniently assumed that all the mass was concentrated at points, and also, that the mass of the bar itself was negligible. But mass in real life is not concentrated at a point, and real bars have a definite mass. So our next stage is to learn how to take this into account. Look at the following illustrations of some simple bodies, and state where you consider the mass centre of each to be located.

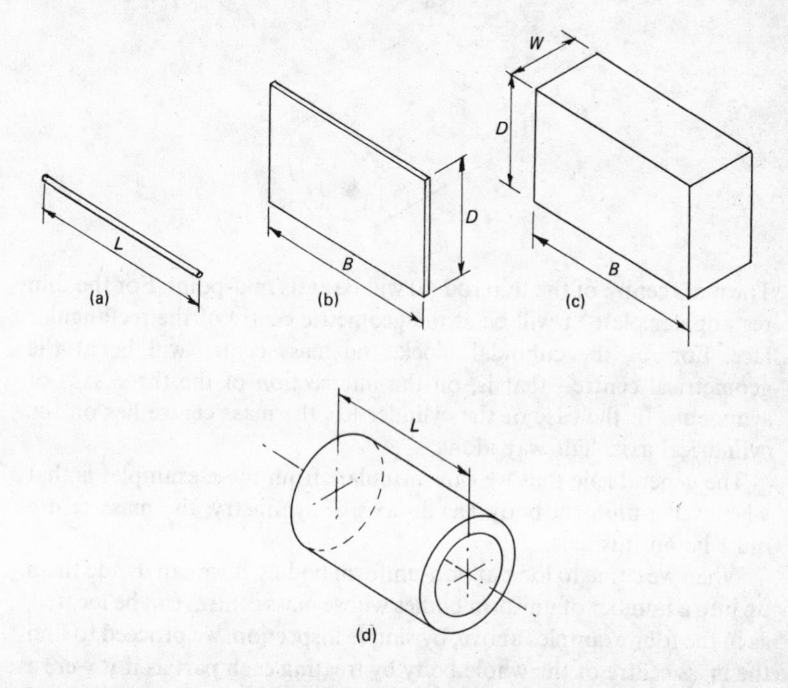

(a) is a thin uniform rod of length L. (b) is a thin uniform plate, of breadth B and depth D. (c) is similar to (b) but the thickness is now not negligible; it has a width W. Finally, (d) is a short length of a hollow cylinder, of length L.

39

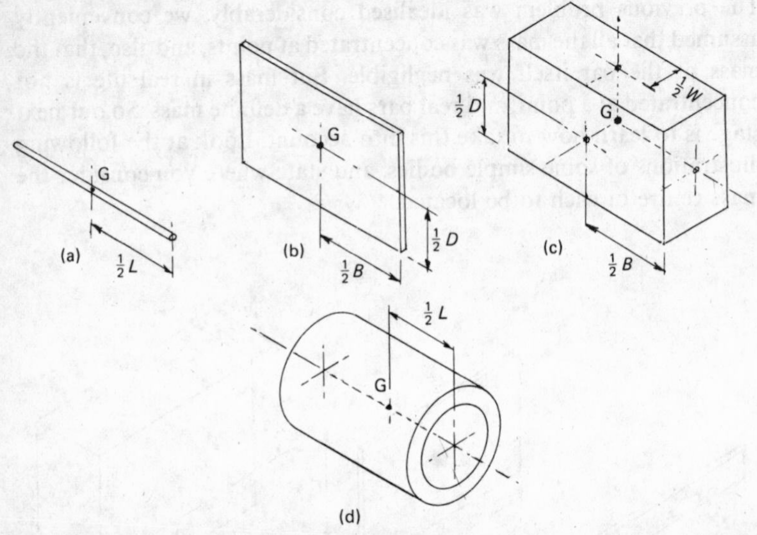

The mass centre of the thin rod (a) will be at its mid-point. For the thin rectangular plate, it will be at the geometric centre of the rectangular face. For (c), the cuboidal block, the mass centre will lie at the geometrical centre—that is, on the intersection of the three axes of symmetry. In the case of the cylinder (d), the mass centre lies on the cylindrical axis, half way along.

The general rule that we can formulate from these examples is that whenever a uniform body has an axis of symmetry, the mass centre must lie on this axis.

When we come to look at non-uniform bodies, if we can divide them up into a number of uniform bodies whose mass centres can be located, as in the four examples above, by simple inspection, we proceed to find the mass centre of the whole body by treating each part as if it were a single mass concentrated at its mass centre. Here is an example. A steel shaft of length $3\frac{1}{2}$ m comprises a section of length 2 m which is 40 mm diameter, coupled to a section of length $1\frac{1}{2}$ m of diameter 50 mm.

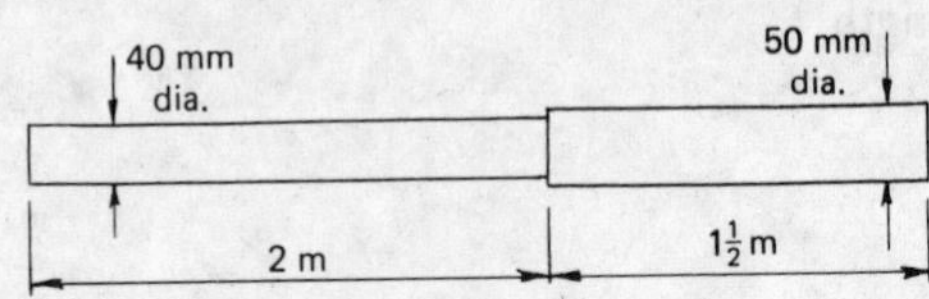

This is equivalent to

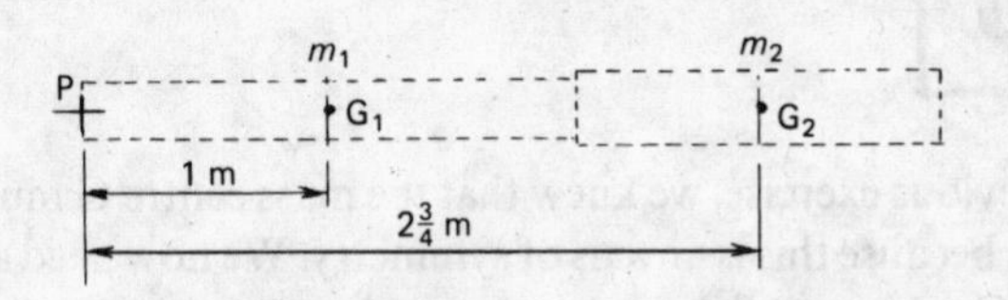

Calling the material density $\rho\,(\mathrm{kg/m^3})$, the masses are

$$m_1 = \frac{\pi}{4}(40^2 \times 10^{-6})2\rho;\ m_2 = \frac{\pi}{4}(50^2 \times 10^{-6})1\tfrac{1}{2}\rho$$

Taking our arbitrary point P at the left-hand end, as shown

$$\bar{x} = \frac{\Sigma(mx)}{\Sigma(m)} = \frac{\frac{\pi}{4}(40^2 \times 10^{-6})2\rho \times 1 + \frac{\pi}{4}(50^2 \times 10^{-6})1\tfrac{1}{2}\rho \times 2\tfrac{3}{4}}{\frac{\pi}{4}(40^2 \times 10^{-6})2\rho + \frac{\pi}{4}(50^2 \times 10^{-6})1\tfrac{1}{2}\rho}$$

Cancelling the common terms $\frac{\pi}{4}$, ρ and 10^{-6}

$$\bar{x} = \frac{40^2 \times 2 \times 1 + 50^2 \times 1\tfrac{1}{2} \times 2\tfrac{3}{4}}{40^2 \times 2 + 50^2 \times 1\tfrac{1}{2}} = \underline{1.944\ \mathrm{m}}$$

and again, you can 'drill' yourself by solving this problem by choosing point P at the right-hand end of the shaft instead of the left.

40

In the previous exercise, we knew that the mass centre G must lie on the shaft axis because this is an axis of symmetry. We now need a method of locating G when we do not have an axis of symmetry. Suppose, for example, that we require to locate the mass centre of a thin metal plate shaped thus

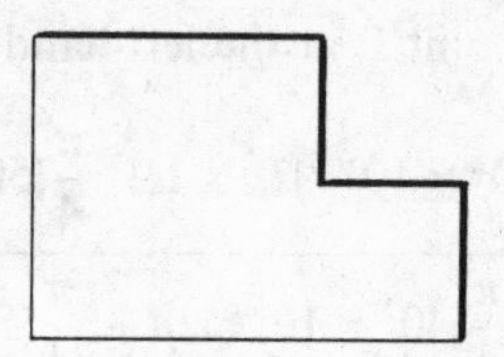

It may help, to begin with, to go back to Frame 35 and to consider the actual weights of the two rectangles making up this body. We know that the centre of gravity of each rectangle will be at its geometric centre. If we choose a point o (at the corner of the plate) and take moments about this point of the two component weights, we can evaluate $\bar{x}$, the distance of the line of action of the resultant weight from o. Thus

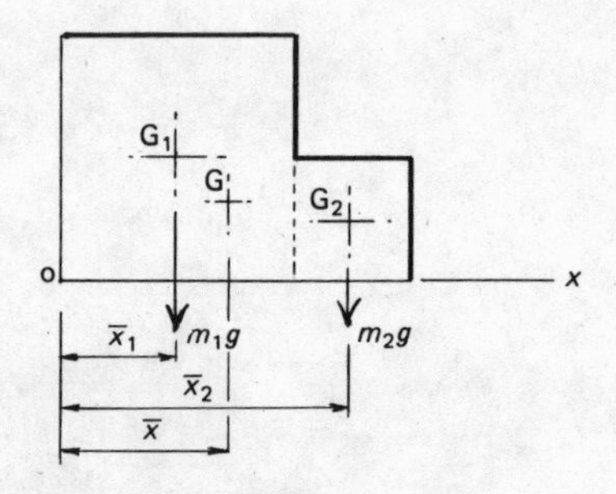

$$mg \times \bar{x} = m_1 g \times \bar{x}_1 + m_2 g \times \bar{x}_2$$

$$\therefore \ \bar{x} = \frac{m_1 \bar{x}_1 + m_2 \bar{x}_2}{m_1 + m_2} = \frac{\Sigma (m\bar{x})}{\Sigma (m)} \quad \text{as before}$$

But of course, this gives us only the distance $\bar{x}$ of the line of action of the resultant weight from point o; it does not actually locate the mass centre G. To do this we have to turn the plate round through 90°

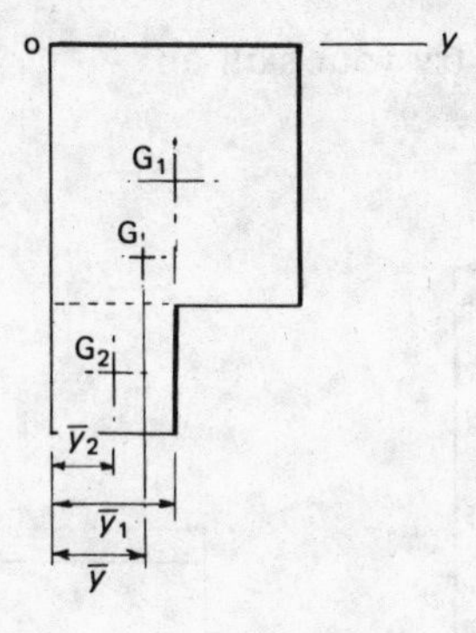

and repeat the process. This procedure will give us the *x and y co-ordinates* of the mass centre. It is not normally necessary actually to redraw the body, as we have done here. We simply show it in a co-ordinate *x–y* system. In this case, the *x* and *y* axes coincide with the two long edges of the plate, but of course the choice of axes is arbitrary.

41

Here is an example to try your skill on.

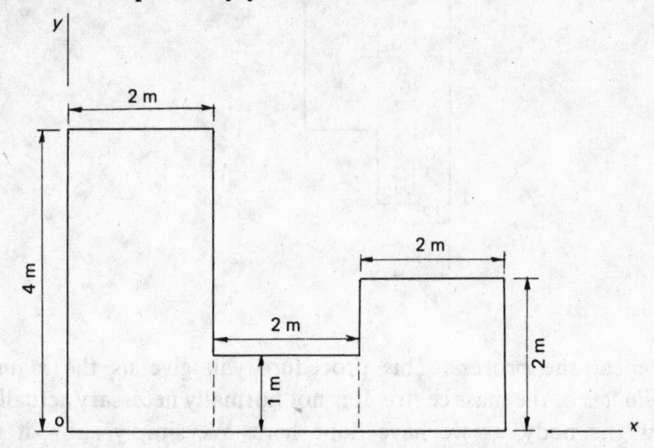

The o–*x*–*y* framework has already been chosen for you; also, the two dotted lines suggest how the plate should be 'split up' for calculation. First of all, calculate $\bar{x}$, the *x* co-ordinate of the mass centre, G. Call the density of the plate material ρ and the thickness *t*. (You should realise by this stage that these terms are going to cancel out.) The value you should get for $\bar{x}$ is 2.429 m. The calculation is done in Frame 42.

42

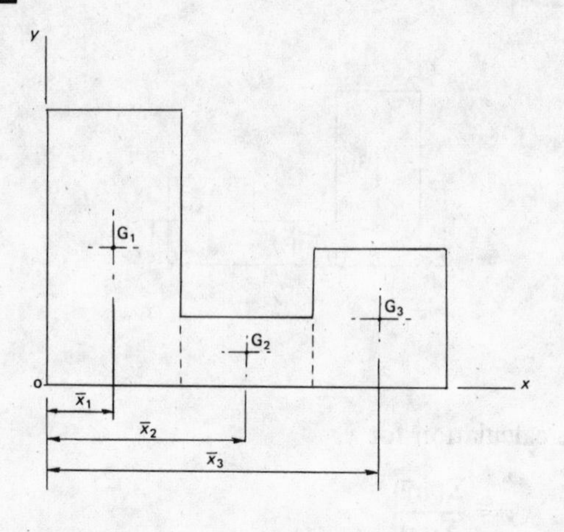

$$\bar{x} = \frac{\Sigma(m\bar{x})}{\Sigma(m)}$$

$$= \frac{\rho t(2 \times 4)1 + \rho t(2 \times 1)3 + \rho t(2 \times 2)5}{\rho t(2 \times 4) + \rho t(2 \times 1) + \rho t(2 \times 2)}$$

$$= \frac{\rho t(8 + 6 + 20)}{\rho t(8 + 2 + 4)}$$

$$= \underline{2.429 \text{ m}}$$

Now repeat the procedure, but this time taking the y co-ordinates of G_1, G_2 and G_3 to obtain $\bar{y}$, the y co-ordinate of G. The answer is 1.5 m.

43

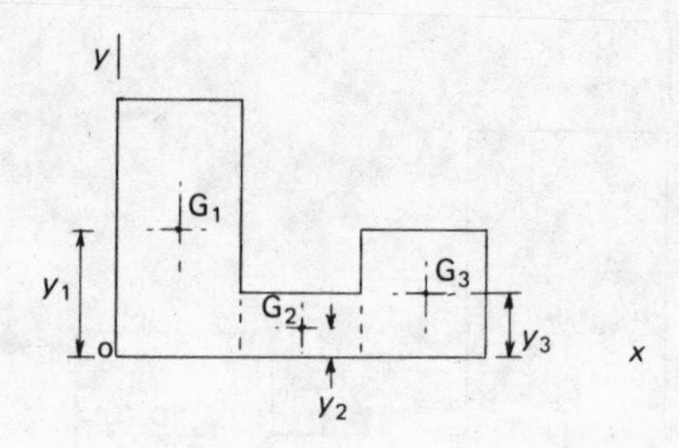

Here is the calculation for y.

$$\bar{y} = \frac{\Sigma(m\bar{y})}{\Sigma(m)}$$

$$= \frac{\rho t(2 \times 4)2 + \rho t(2 \times 1)\tfrac{1}{2} + \rho t(2 \times 2)1}{\rho t(2 \times 4) + \rho t(2 \times 1) + \rho t(2 \times 2)}$$

$$= \frac{16 + 1 + 4}{8 + 2 + 4}$$

$$= \underline{1.5 \text{ m}}$$

An interesting observation resulting from this calculation is that the mass centre of a body does not necessarily lie within the material of the body itself.

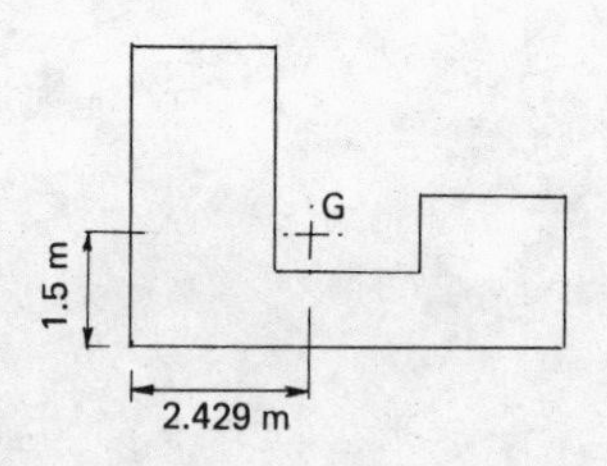

You can now give yourself as much exercise as you need with this one example. You can divide the figure into a different arrangement of rectangles. For example

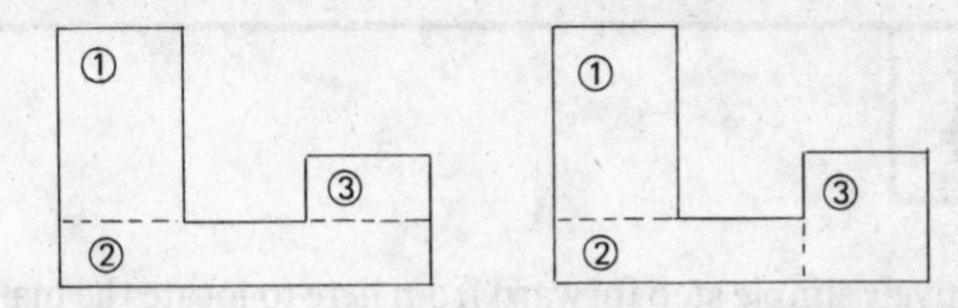

and you should still get the same values for $\overline{x}$ and $\overline{y}$. Or again, you can choose a different set of x–y axes. If you do, you will not get the same values of $\overline{x}$ and $\overline{y}$, but the actual position of G within the figure should, of course, be the same. One thing you must always remember: values of x and y may be positive or negative. Suppose, for example, you solve this problem using the division into rectangles suggested in the left-hand sketch above, and that you choose x and y axes thus

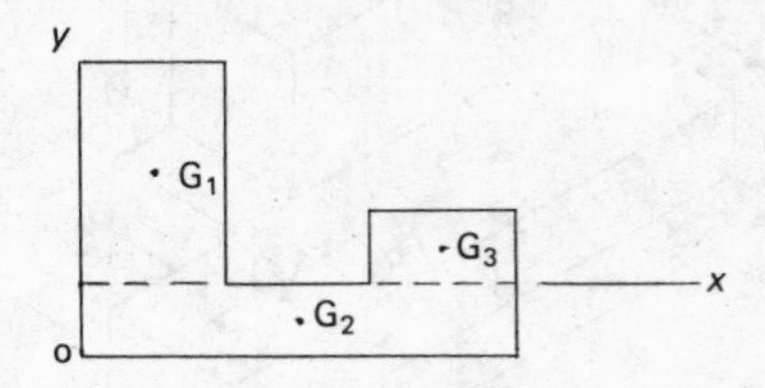

In calculating $\overline{y}$, you must put in a value of *minus* 0.5 m for the y co-ordinate of G_2, because it lies on the other side of the x-axis.

44

It is a relatively simple step forward from here to locate the mass centre of a solid irregular body with no axes of symmetry.

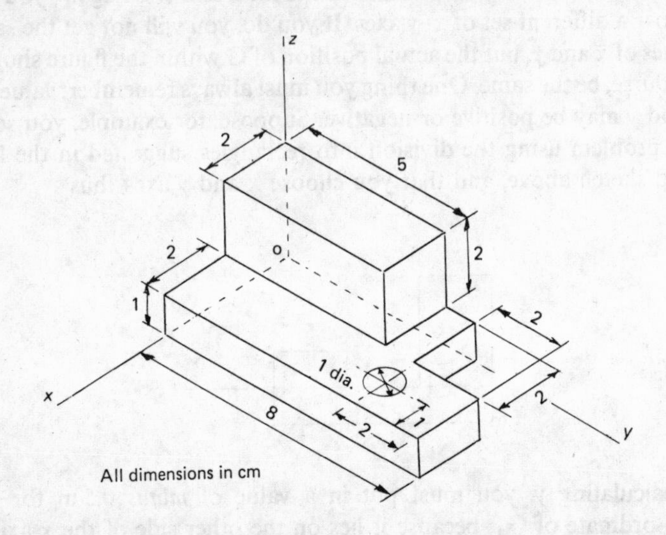

Find the three co-ordinates, $\bar{x}$, $\bar{y}$ and $\bar{z}$ of the mass centre of the body shown above, with reference to the x–y–z axes shown. These axes coincide with the three 'rear edges' of the body. A hole in a solid body can be treated as a negative mass. The 2 cm by 2 cm cutaway can be treated in the same manner, and this is how it is done in the solution given in the following frame, but there are other ways of subdividing the body. Four subdivisions will be required, and in order that you can check your work, you may prefer to use the method of subdivision used in the solution given. This is (1) the 2 cm by 2 cm by 5 cm rectangular prism on the top; (2) the 4 cm by 1 cm by 8 cm rectangular prismatic base; (3) the 'negative' 2 cm by 2 cm cutaway on the corner; (4) the 'negative' 1 cm diameter hole. You should expect by now that the density of ρ of the material will cancel out, and it has not been included in the following solution. To save you anxiety, all co-ordinates of all the submasses will be positive, using the axes shown.

45

$$\bar{x} = \frac{(2 \times 5 \times 2)1 + (4 \times 8 \times 1)2 - (2 \times 2 \times 1)1 - (\pi/4 \times 1^2 \times 1)3}{(2 \times 5 \times 2) + (4 \times 8 \times 1) - (2 \times 2 \times 1) - (\pi/4 \times 1^2 \times 1)}$$

$$= \frac{27.644}{47.215} = \underline{1.644 \text{ cm}}$$

$$\bar{y} = \frac{(2 \times 5 \times 2)2\tfrac{1}{2} + (4 \times 8 \times 1)4 - (2 \times 2 \times 1)7 - (\pi/4 \times 1^2 \times 1)6}{(2 \times 5 \times 2) + (4 \times 8 \times 1) - (2 \times 2 \times 1) - (\pi/4 \times 1^2 \times 1)}$$

$$= \frac{145.288}{47.215} = \underline{3.077 \text{ cm}}$$

$$\bar{z} = \frac{(2 \times 5 \times 2)2 \times (4 \times 8 \times 1)\tfrac{1}{2} - (2 \times 2 \times 1)\tfrac{1}{2} - (\pi/4 \times 1^2 \times 1)\tfrac{1}{2}}{(2 \times 5 \times 2) + (4 \times 8 \times 1) - (2 \times 2 \times 1) - (\pi/4 \times 1^2 \times 1)}$$

$$= \frac{53.607}{47.215} = \underline{1.138 \text{ cm}}$$

46

Now we have to learn how to locate the mass centre of a body which cannot be subdivided into circles and rectangles and other very simple shapes. How, for example, do we find the position of the mass centre of a thin flat semicircular plate? In such cases, we have to resort to the use of the calculus, and if you have not learned how to use this, you will have to skip this frame. Here is the plate, shown with a set of co-ordinate axes, o–x–y, with o being the centre of the semicircle.

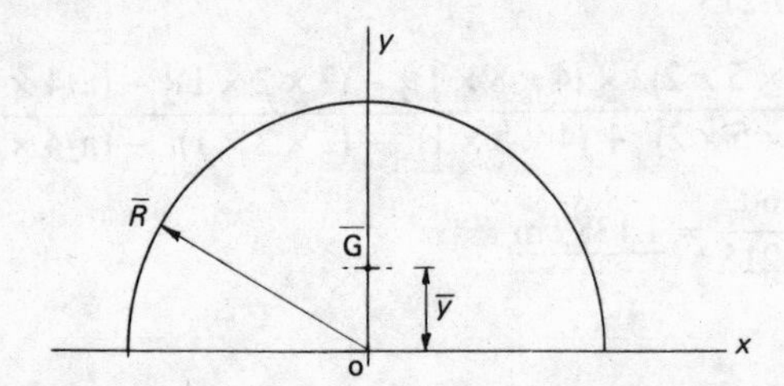

We know that the mass centre must lie on the axis of symmetry, which is the y-axis, so we have to determine $\bar{y}$ only. Consider an 'increment' at distance y from the x-axis, of thickness δy.

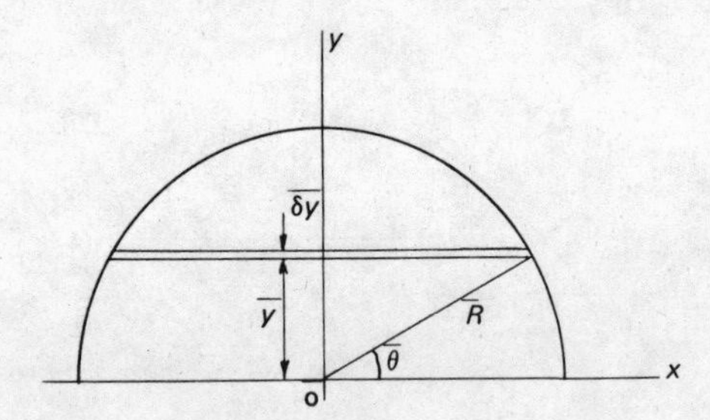

If the length of this strip is b, its mass will be $\rho t \times b \times \delta y$ where ρ and t are respectively the density of the plate material and the thickness. The mass of the whole plate will be $\rho t \times \frac{1}{2}\pi R^2$. With circular functions, it is usually more convenient to have a variable θ than a variable y (which would lead to some awkward integrals, with 'substitutions' required). From the figure above, we can see that

$$y = R\sin\theta$$

Hence

$$\frac{dy}{d\theta} = R\cos\theta$$

Also

$$b = 2 \times R\cos\theta$$

The 'limits' of θ will be from zero to $\frac{1}{2}\pi$ (that is, 90°)

$$\overline{y} = \frac{\Sigma(\delta m \times y)}{\Sigma(\delta m)} = \frac{\Sigma(\rho t(2R\cos\theta)\,(R\cos\theta\, d\theta)\,(R\sin\theta))}{\rho t \times \frac{1}{2}\pi R^2}$$

$$= \frac{\rho t \times \int_0^{\pi/2} 2R^3 \sin\theta \cos^2\theta\, d\theta}{\rho t \times \frac{1}{2}\pi R^2}$$

$$= \frac{4R}{\pi}\int_0^{\pi/2} (-\cos^2\theta)\, d(\cos\theta)$$

$$= \frac{4R}{\pi}\left[-\tfrac{1}{3}\cos^3\theta\right]_0^{\pi/2}$$

$$= \frac{4R}{\pi}\left\{(-\tfrac{1}{3}\cos^3\pi/2) - (-\tfrac{1}{3}\cos^3 0)\right\}$$

$$= \underline{\frac{4R}{3\pi}}$$

47

The work of Frame 46 is verging outside the area of 'Elementary Statics' and you should not be too concerned if you find it a bit too much for you at this stage. But even if you cannot follow the working, you can make use of the result to extend your capacity to calculate the position of the mass centre of an irregular body; you should now be able to deal with semicircular parts and semicircular holes or cutaways. While on the same topic, what about the mass centre of a thin flat triangular sheet? We can locate this without actually using the calculus, and you may have done this previously yourself. If we mentally divide a triangular sheet into a number of thin strips parallel to the base, we know that the mass centre of each strip will be at its mid-point.

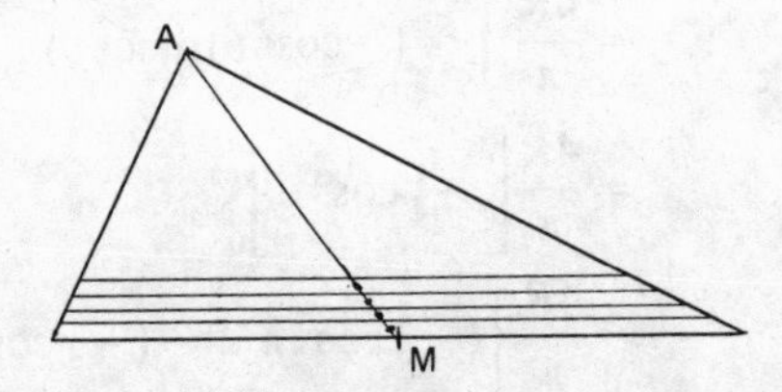

The mass centres of all these strips lie on a straight line joining the apex A to the mid-point of the base, M. This line is called a *median.* So the mass centre of the whole plate lies somewhere on the median. If we now turn our triangle round and make another edge the 'base', we can draw a second median, and the same reasoning will apply. So the mass centre of the triangular plate must lie on the intersection of the medians. Some fairly elementary geometry would be needed to prove (which we shall not do here) that the medians of a triangle intersect at a point one-third the way along the length. We can now locate our mass centre.

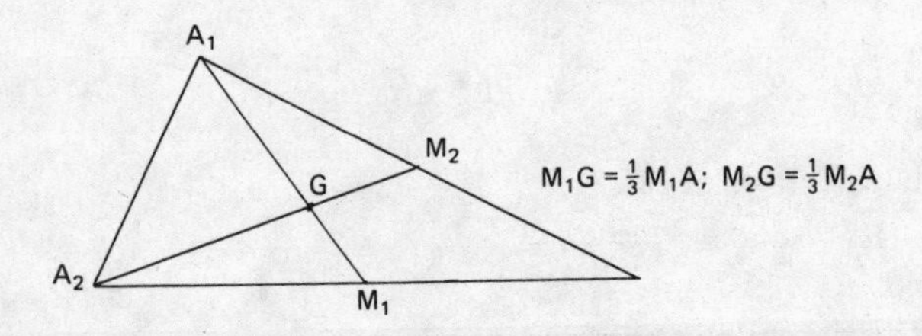

Now see if you can make use of this new information to find the x co-ordinate ($\bar{x}$) of the thin plate shown below. Use the method employed in Frame 40 together with some very elementary geometry of similar triangles. Look at the next frame for the solution.

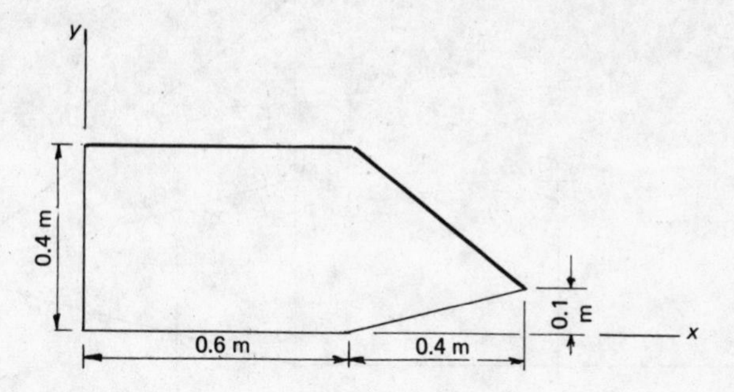

48

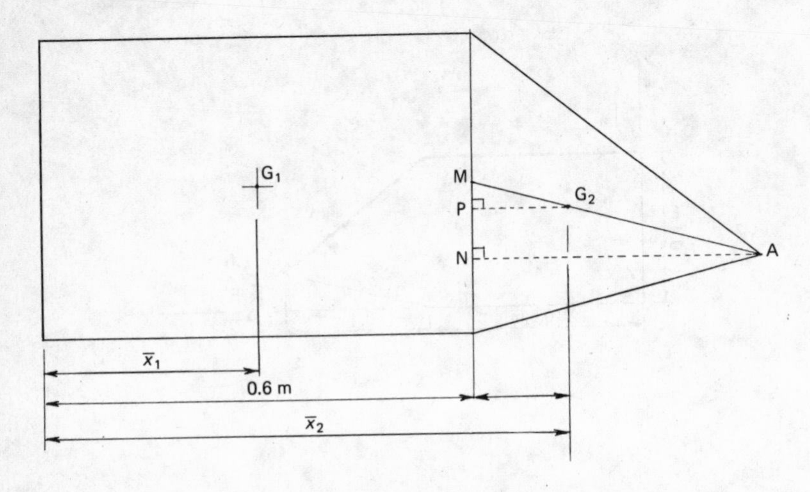

$$\bar{x} = \frac{\Sigma(mx)}{\Sigma(m)} = \frac{m_1 x_1 + m_2 x_2}{m_1 + m_2}$$

The distance $\bar{x}_1$ for the rectangle is 0.3 m (half the width). Construct the perpendiculars, AN and G_2P. It can then be seen that $\bar{x}_2 = 0.6$ m plus the length of perpendicular G_2P. We see two similar triangles, MG_2P and MAN. We know that MG_2 is one-third of MA. Therefore G_2P will be one-third of AN and AN is 0.4 m.

$$\therefore\ \bar{x} = \frac{\rho t(0.4 \times 0.6)0.3 + \rho t(\frac{1}{2} \times 0.4 \times 0.4)(0.6 + \frac{1}{3} \times 0.4)}{\rho t(0.4 \times 0.6) + \rho t(\frac{1}{2} \times 0.4 \times 0.4)}$$

$$= \frac{0.072 + 0.08 \times 0.6333}{0.24 + 0.08}$$

$$= \underline{0.3833 \text{ m}}$$

You should not find the geometry too difficult to determine the value of $\bar{y}$, although it is not quite so simple as that of the above calculation. A brief solution follows, in case you have difficulty.

49

$$\bar{y} = \frac{\rho t(0.4 \times 0.6)\,0.2 + \rho t(\frac{1}{2} \times 0.4 \times 0.4)(0.1 + \frac{2}{3} \times 0.1)}{\rho t(0.4 \times 0.6) + \rho t(\frac{1}{2} \times 0.4 \times 0.4)}$$

$$= \underline{0.1917 \text{ m}}$$

Using calculus, it is possible to show that the mass centre of a solid uniform right cone is located at a distance of one-quarter of its height from the base (and, of course, on its central axis). We shall not prove this, but we can use the result to find the position of the mass centre of a truncated cone, such as a tapered shaft.

50

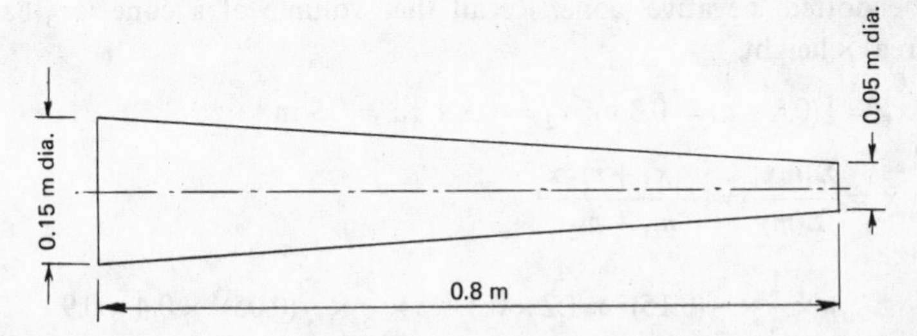

Make use of the information given at the end of the previous frame to calculate the position of the mass centre of the tapered shaft shown above. Treat it as two complete cones, the second being a 'negative' cone. Before you begin your calculations, it is often a good plan to have a guess at the answer. In this case, if the shaft were the same length but of constant diameter, instead of being tapered, we know that the mass centre would be in the middle—that is, 0.4 m from the left-hand end. Tapering means that there will be more mass to the left of centre than to the right; the mass centre will thereby be shifted somewhat to the left of centre. Your answer, therefore, must be *less* than 0.4 m. If it turns out to be more, look back to find where your mistake was. Furthermore, if the shaft were of the same length and tapered to a point at the right-hand end, then we know that the mass centre would then be $\frac{1}{4} \times 0.8 = 0.2$ m from the left-hand end. The shaft as it is means that there is more mass to the right of centre than there would have been if it had tapered to a point; this will shift the mass centre a little to the right of 0.2 m. Our answer, therefore, must lie between 0.2 m and 0.4 m.

51

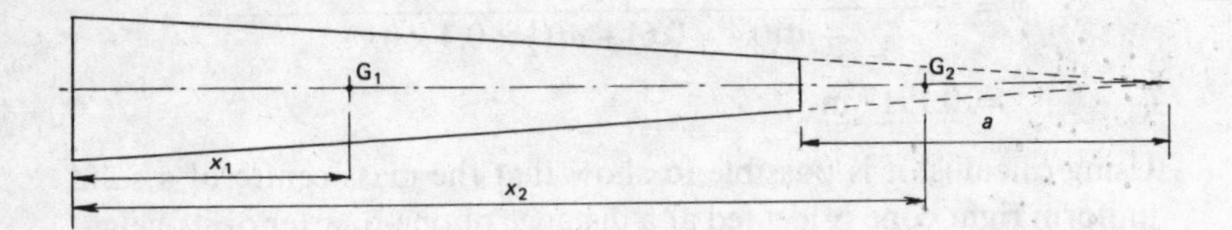

The sides will converge together at distance a from the right-hand end. From similar triangles

$$\frac{a}{0.05} = \frac{a+0.8}{0.15} \quad \text{giving } a = 0.4 \text{ m}$$

G_1 is the mass centre of the whole cone; G_2 is the mass centre of the dotted 'negative' cone. Recall that volume of a cone is $\frac{1}{3}$(base area) × height.

$$x_1 = \tfrac{1}{4}(0.8+a) = 0.3 \text{ m}; \; x_2 = 0.8 + \tfrac{1}{4}a = 0.9 \text{ m}$$

$$\bar{x} = \frac{\Sigma(mx)}{\Sigma(m)} = \frac{m_1x_1 + m_2x_2}{m_1 + m_2}$$

$$= \frac{\rho \times \frac{1}{3} \times \frac{\pi}{4}(0.15)^2 \times 1.2 \times 0.3 - \rho \times \frac{1}{3} \times \frac{\pi}{4}(0.05)^2 \times 0.4 \times 0.9}{\rho \times \frac{1}{3} \times \frac{\pi}{4}(0.15)^2 \times 1.2 - \rho \frac{1}{3} \times \frac{\pi}{4}(0.05)^2 \times 0.4}$$

$$\text{Cancelling } \bar{x} = \frac{0.15^2 \times 1.2 \times 0.3 - 0.05^2 \times 0.4 \times 0.9}{0.15^2 \times 1.2 - 0.05^2 \times 0.4}$$

$$= \frac{0.0081 - 0.0009}{0.027 - 0.001} = \underline{0.2769 \text{ m}}$$

52

Now if you look back at all the examples concerning the mass centre of thin flat plates—Frames 42, 43, 46, 48 and 49—you will have noticed that the density ρ and the thickness t have always cancelled out in the expressions for $\bar{x}$ and $\bar{y}$. We shall re-examine the problem by considering the mass centre of any flat plate, in general terms, instead of a specific problem.

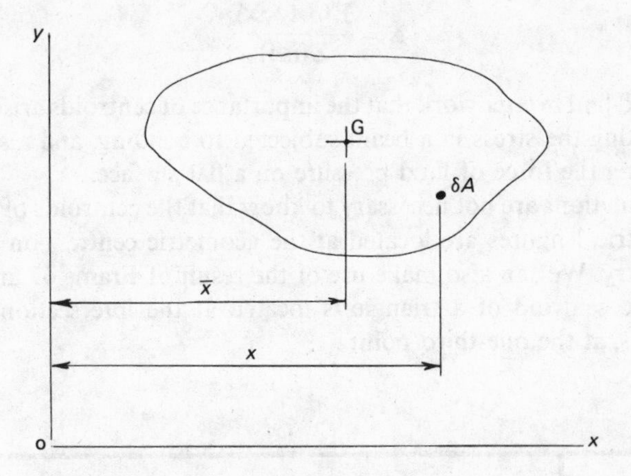

Again, call the plate material density ρ and the thickness t. Consider an incremental mass of surface area δA. The mass of this 'element' is $\rho t \times \delta A$.

$$\bar{x} = \frac{\Sigma(\delta m \times x)}{\Sigma(\delta m)} = \frac{\Sigma(\rho t \times \delta A \times x)}{\Sigma(\rho t \times \delta A)} = \frac{\Sigma(\delta A \times x)}{\Sigma(\delta A)}$$

In Frame 37 we found that the position of the 'centre of gravity' was independent of the actual weight of a body, and for this reason, we adopted the more general term 'mass centre'. Now we see that in the case of a thin flat lamina or plate, the position of the mass centre is actually independent of the mass! It has become only a function of the shape of the plate. The quantity

$$\bar{x} = \frac{\Sigma(\delta A \times x)}{\Sigma(\delta A)}$$

can now be evaluated for any prescribed *area* and may have nothing at all to do with force, weight or mass. It can be merely a geometric quantity, or property, of a closed figure. In such a case, it is unrealistic to speak of 'centre of gravity' or 'centre of mass', and instead we use the term '*centroid*'. If you need to define 'centroid', you may say that it is the point on a surface at which the resultant force would act if the surface were subjected to a uniformly distributed force over the whole surface. But it is unsatisfactory to have to use the concept of force to define a merely geometric property of a figure, and it is better to define the centroid merely in terms of the expression above. Thus

"The centroid of a closed figure is a point, distant $\bar{x}$ from an arbitrary axis such that

$$\bar{x} = \frac{\Sigma(\delta A \times x)}{\Sigma(\delta A)} \text{"}$$

You will find in later work that the importance of centroids arises when calculating the stress in a beam subjected to bending; and also when analysing the force of fluid pressure on a flat surface.

Calculations are not necessary to know that the centroids of regular symmetrical figures are located at the geometric centre—on axes of symmetry. We can also make use of the result of Frame 47 and state that the centroid of a triangle is located at the intersection of the medians, at the one-third point.

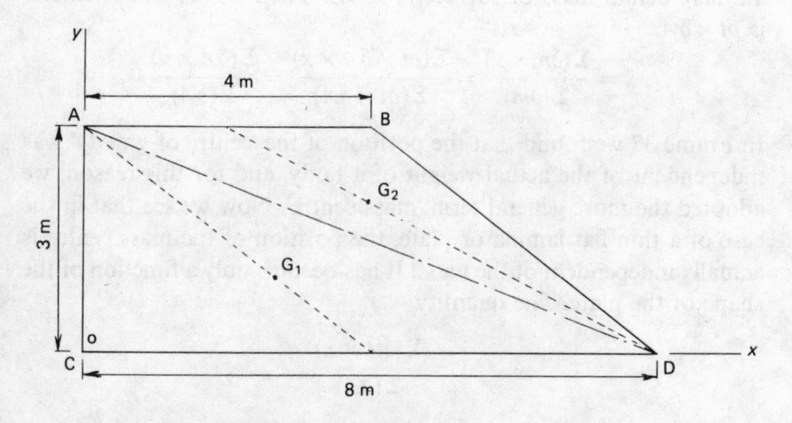

It is quite clear from the contents of the previous frame that the calculation for the locations of centroids is very similar to that for locations of centres of mass. Determine, therefore, the values of $\bar{x}$ and $\bar{y}$ for the trapezium shown here. The figure has already been divided into two triangles for you, and the medians drawn. You may need help by looking back to Frame 48. You should obtain answers of $\bar{x} = 3\frac{1}{9}$ m and $\bar{y} = 1\frac{1}{3}$ m. Frame 54 shows the working.

54

$$\bar{x} = \frac{A_1\bar{x}_1 + A_2\bar{x}_2}{A_1 + A_2}$$

$$= \frac{(\frac{1}{2} \times 8 \times 3) \times \frac{2}{3} \times 4 + (\frac{1}{2} \times 4 \times 3) \times (2 + \frac{1}{3} \times 6)}{(\frac{1}{2} \times 8 \times 3) + (\frac{1}{2} \times 4 \times 3)}$$

$$= \frac{32 + 24}{12 + 6}$$

$$= \underline{3\frac{1}{9}\text{ m}}$$

$$\bar{y} = \frac{(\frac{1}{2} \times 8 \times 3) \times \frac{1}{3} \times 3 + (\frac{1}{2} \times 4 \times 3) \times \frac{2}{3} \times 3}{12 + 6}$$

$$= \frac{12 + 12}{12 + 6}$$

$$= \underline{1\frac{1}{3}\text{ m}}$$

55

We shall bring the theoretical work of this programme to a close with a consideration of the properties of an area bounded by a parabola. The following sketch shows a parabola of the general form $y = kx^2$.

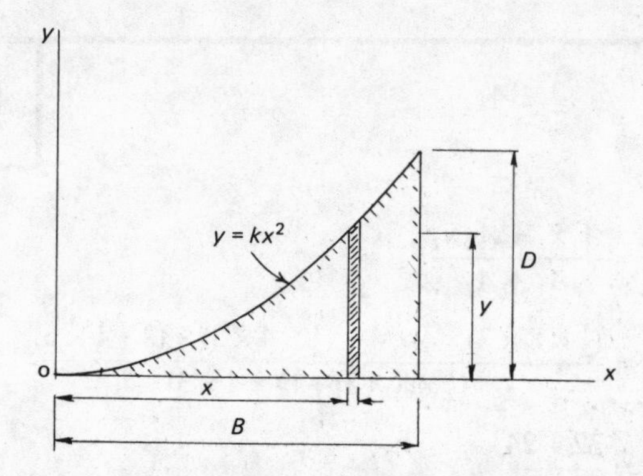

The vertex of such a parabola will be at the origin, o. We begin by calculating the area under the curve, shown hatched, from $x = 0$ to $x = B$. At this point, the 'height' of the area is D. The area $= \Sigma(\delta A) = \Sigma(y \times \delta x) = \Sigma(kx^2 \times \delta x)$. Integrate and insert the limits. You should find that $A = \frac{1}{3}BD$. See the next frame.

56

$$A = \Sigma(\delta A) = \Sigma(y \times \delta x) = \Sigma(kx^2 \times \delta x)$$

$$= \int_0^B kx^2\,\mathrm{d}x$$

$$= [\tfrac{1}{3}kx^3]_0^B$$

$$= \tfrac{1}{3}kB^3$$

k can be eliminated from this expression: when $x = B$, $y = D$. Thus

$$y = kx^2$$

$$\therefore \quad D = kB^2$$

$$\therefore \quad k = \frac{D}{B^2}$$

$$\therefore \quad A = \frac{1}{3}kB^3 = \frac{1}{3}\left(\frac{D}{B^2}\right)B^3 = \underline{\frac{1}{3}BD}$$

Now follow the usual procedure and determine $\bar{x}$, the distance of the centroid from the y-axis. The answer is $\frac{3}{4}B$. Working is in the next frame.

57

$$\bar{x} = \frac{\Sigma(\delta A \times x)}{\Sigma(\delta A)} = \frac{\Sigma(y \times \delta x \times x)}{A}$$

$$= \int_0^B \frac{(kx^2 \times x)\,dx}{\frac{1}{3}BD}$$

$$= \frac{k[\frac{1}{4}x^4]_0^B}{\frac{1}{3}BD}$$

$$= \frac{(D/B^2)(\frac{1}{4}B^4)}{\frac{1}{3}BD}$$

$$= \underline{\tfrac{3}{4}B}$$

Now let us complete the rectangle and examine the shape that is left when the area under the parabola is removed.

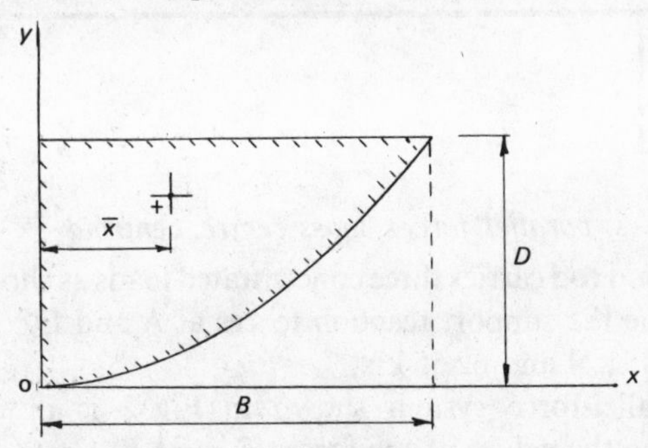

Find the value of $\bar{x}$ for this figure. No need to integrate. Treat the shaded area as a rectangle $B \times D$ and a 'negative' parabola. Then

$$A\bar{x} = A_1\bar{x}_1 - A_2\bar{x}_2$$

from which you should be able to show that $\bar{x} = \frac{3}{8}B$.

58

The area A is the area of the rectangle minus the area of the parabola just determined.

$$A = BD - \tfrac{1}{3}BD = \tfrac{2}{3}BD$$

$$A\bar{x} = A_1\bar{x}_1 - A_2\bar{x}_2$$

$$\therefore \quad \bar{x} = \frac{(BD)\tfrac{1}{2}B - (\tfrac{1}{3}BD)\tfrac{3}{4}B}{\tfrac{2}{3}BD}$$

$$= \frac{\tfrac{1}{2}B^2D - \tfrac{1}{4}B^2D}{\tfrac{2}{3}BD}$$

$$= \underline{\tfrac{3}{8}B}$$

The properties of these two parabolic shapes are extremely useful when the deflections of beams are calculated using the method of Moment–Area. It is very important to remember that the values that we have determined apply *only* to parabolic shapes in which the vertex is located at the origin o.

This completes the work of this programme. Some 'drill' examples follow in Frame 59, covering the work since Frame 20, and this is followed by a few general revision examples, in Frame 60, covering the whole programme.

59

'Drill' exercises: parallel forces, mass centre, centroid

1. A light rigid rod carries three concentrated loads as shown in Fig. 1. Determine the support reaction forces at A and D.
   Ans. 25.45 kN and 62.55 kN.
2. The parallel-force system shown in Fig. 2 is in equilibrium. Determine the values of the forces F_1 and F_2.

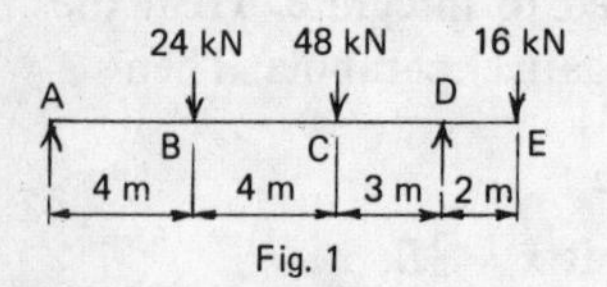

Fig. 1

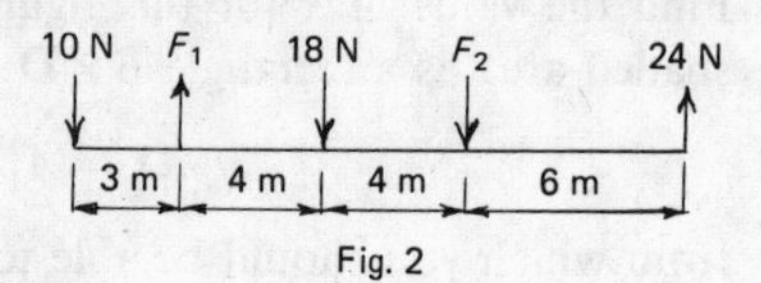

Fig. 2

3. Determine the resultant of the parallel-force system shown in Fig. 3 and the location of its line of action.
Ans. 80 kN, 5 m from left-hand end.
4. A steel beam of uniform section has a mass per unit length of 4 kg/m. It carries concentrated loads as shown in Fig. 4 and is supported by a prop at A and a steel wire hanger at C. Calculate the forces in the prop and the hanger.
Ans. 21.72 kg wt (213.1 N) and 50.28 kg wt (493.2 N).

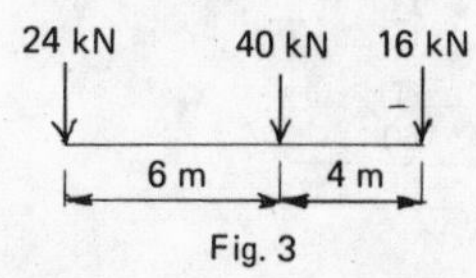

Fig. 3

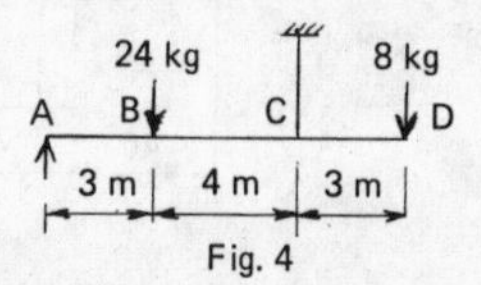

Fig. 4

5. Calculate the magnitude, direction and point of location of the resultant of the eight forces shown in Fig. 5, acting at the corners of the square ABCD, which measures 4 m by 4 m.
Ans. 20.81 N at 35.2° to right of vertical, passing through a point 1.41 m to right of AD and 1.33 m above CD.
6. ABCD is a rectangle, AB = 6 m and BC = 4 m. E lies on AB; AE = 4 m. F lies on BC; BF = 2 m. G lies on CD; CG = 2 m. Forces of 10 N, 12 N and 8 N act at A in the respective directions AB, AF, AG. A single force of 6 N acts at B in the direction GB. A single force of 8 N acts at C in the direction EC. Forces of 2 N and 4 N act at G in the respective directions GE and GC. A single force of 10 N acts at D in the direction DC. Determine the magnitude and direction of the resultant of the eight forces and calculate where its line of action passes through CD.
Ans. 48.19 N at 11.05° below horizontal. Intersects DC produced 13.17 m from D.
7. Locate the mass centre of the light rod carrying the four concentrated masses shown in Fig. 6.
Ans. 3.667 m from left-hand end.

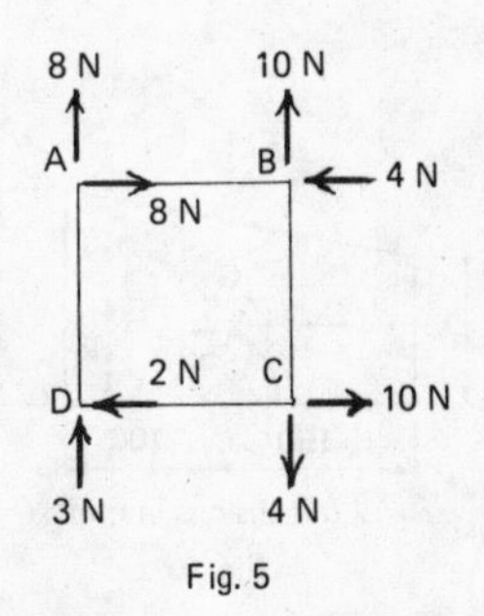

Fig. 5

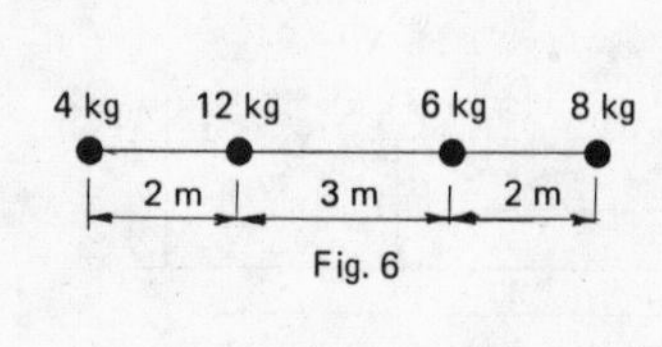

Fig. 6

8. A steel shaft has the dimensions shown in Fig. 7. Determine the location of the mass centre.
Ans. 83.96 cm from left-hand end.

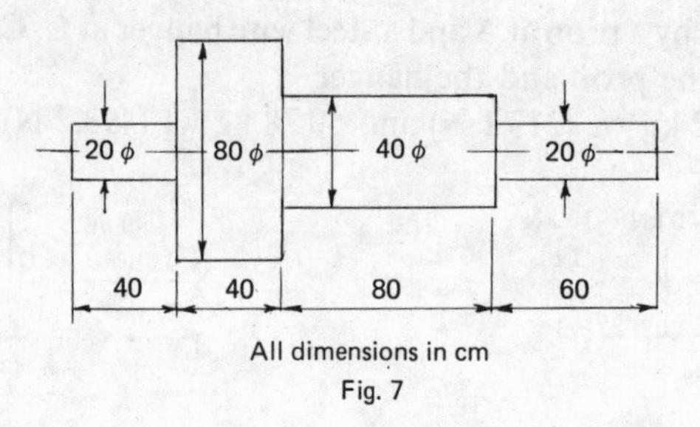

Fig. 7

9. A steel shaft 1.5 m long has a circular cross-section tapering uniformly from 20 mm diameter at one end to 50 mm diameter at the other end. Calculate the distance of the mass centre from the smaller end.
Ans. 0.952 m.
10. Prove that the mass centre of a uniform solid hemisphere is located at a distance of $\frac{3}{8}$ of the radius from the centre.
11. Prove that the mass centre of a thin uniform hemispherical shell is located at a distance of $\frac{1}{2}$ the radius from the centre.
12. Determine the co-ordinates $\bar{x}$ and $\bar{y}$ of G, the centroid of the figure shown in Fig. 8.
Ans. $\bar{x}$ = 35.56 cm; $\bar{y}$ = 17.78 cm.
13. Locate the centroid of the figure shown in Fig. 9. Assume the centroid of a semicircle to be a distance of $4R/3\pi$ from the centre.
Ans. $\bar{x}$ = 130.94 mm; $\bar{y}$ = 69.39 mm.

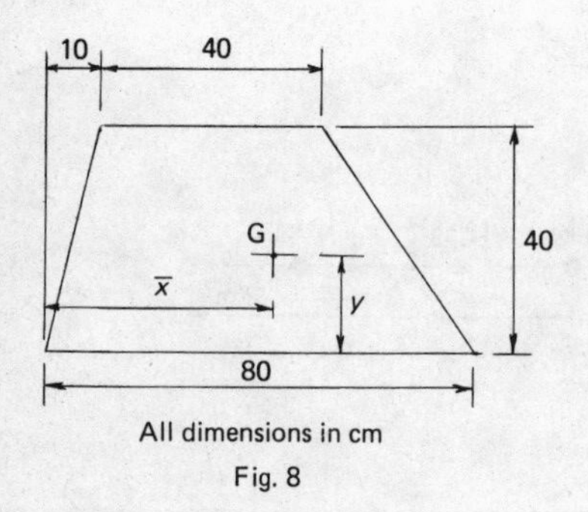

Fig. 8

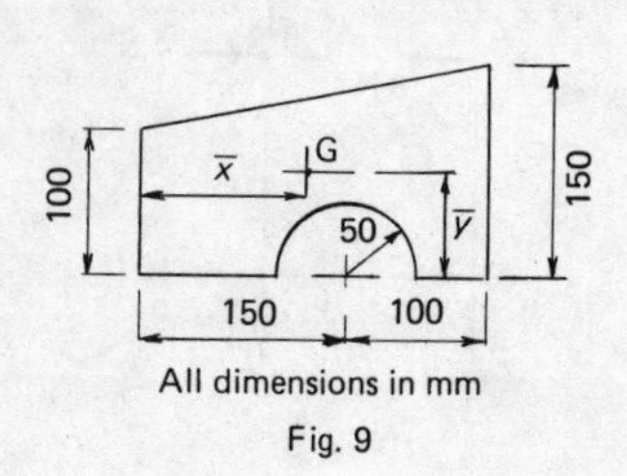

Fig. 9

14. Prove that the centroid of a half-ellipse of major axis $2a$ and minor axis $2b$ is located at a distance of $4a/3\pi$ from the centre. The equation for the ellipse is

$$\frac{x^2}{a^2}+\frac{y^2}{b^2}=0$$

and the substitution $x = a\sin\theta$ and $y = b\cos\theta$ will be found useful.

60

Revision examples

1. (a) Given that the force system shown in Fig. 1 is in equilibrium, determine the values of the forces F_1 and F_2.
   (b) Given that $F_1 = F_2 = 0$, determine the magnitude and direction of the resultant of the system.
   Use a graphical method.
   Ans. See Problem 2.
2. Solve Problem 1 analytically.
   Hint: to determine F_1, resolve all forces in directions parallel to, and perpendicular to, F_2. Write the equation of equilibrium along the direction perpendicular to F_2. Determine F_2 similarly.
   Ans. (a) $F_1 = 4.956$ N; $F_2 = 1.587$ N. (b) $R = 4.249$ N at 32.19° below horizontal.
3. A uniform beam AB has a mass of 450 kg and is 2.5 m long. It supports a hanging mass of 65 kg at the end B and is hinged to a fixed support at A. A light steel wire is connected at C, 1.1 m from A. This passes over a small fixed pulley at D which is 1.4 m vertically above A, and from there to a winding drum. Figure 2 shows the arrangement. Calculate the force required in the wire to raise the loaded beam (a) when it is horizontal, as shown, (b) when it has been raised 30° above the horizontal.
   Hint: for solution to (b), solve the triangle ACD using cosine and sine formulae.
   Ans. (a) 8.22 kN. (b) 5.898 kN.

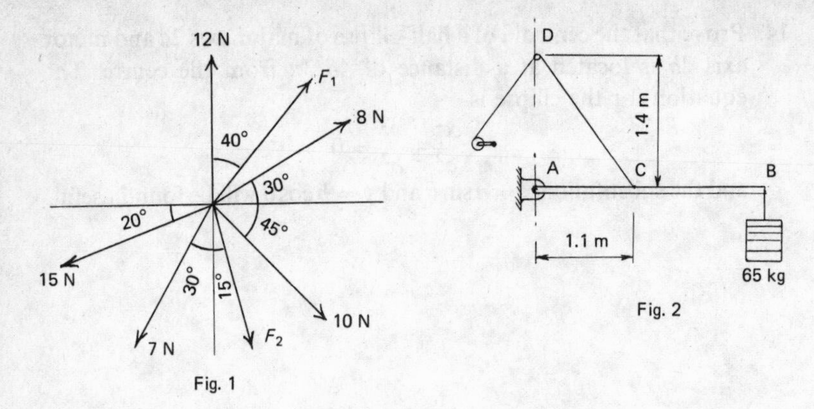

Fig. 1

Fig. 2

4. (a) For the system of parallel forces shown in Fig. 3, if $P = 4$ kN and $Q = 6$ kN, find the magnitude and the line of action of the resultant force.
(b) If the whole system is in equilibrium, determine values of P and Q.
Ans. (a) 14 kN down at $4\frac{3}{7}$ m from left-hand end. (b) $12\frac{1}{3}$ kN; $11\frac{2}{3}$ kN.

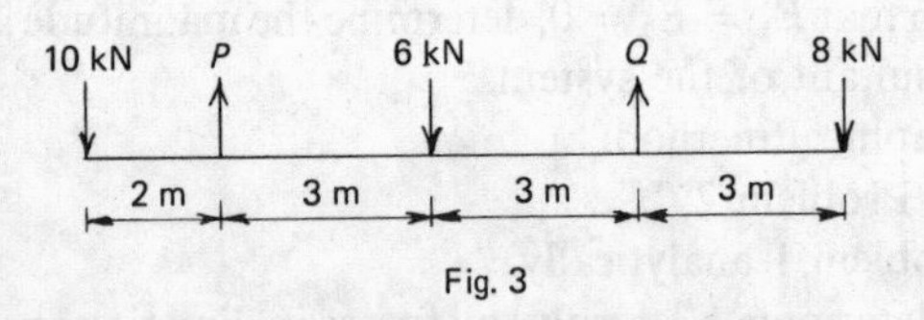

Fig. 3

5. ABCDEF is a regular hexagon of length of side a. Forces act at A of 10 N in the direction AB, 8 N in the direction AC, 5 N in the direction of AD and 4 N in the direction of AE. A force of 6 N acts at B in the direction of CB. A force of 8 N acts at C in the direction of CF. A force of 12 N acts at D in the direction CD. A force of 7 N acts at E in the direction ED. A force of 10 N acts at F in the direction EF. Determine the magnitude and direction of the resultant force of the system, and state where it crosses line DE.
Ans. 9.910 N 63.46° below horizontal; 3.014a from E on ED produced.
6. Determine the three co-ordinates $\bar{x}$, $\bar{y}$ and $\bar{z}$ for the component shown in Fig. 4. The 1.2 cm diameter hole in the stepped portion passes through the full depth of the step.
Ans. $\bar{x} = 1.457$ cm; $\bar{y} = 1.194$ cm; $\bar{z} = 0.883$ cm.

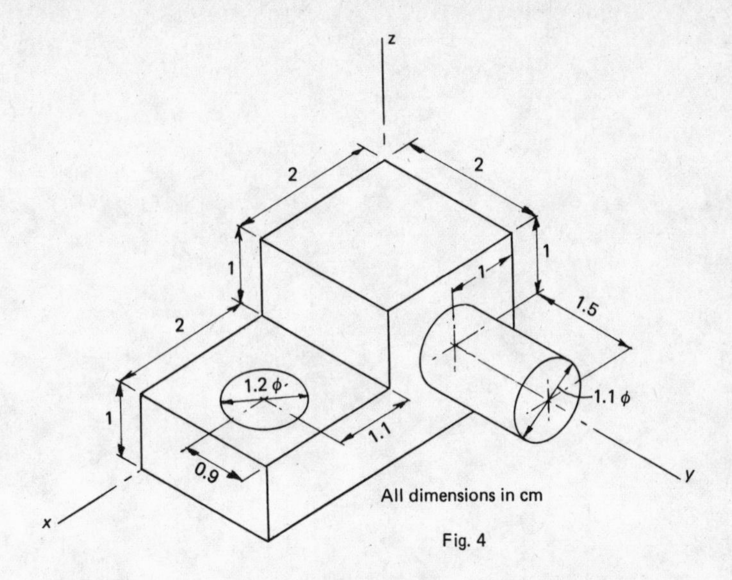

Fig. 4

7. A concrete column has the form of a truncated cone with a cylindrical hole along its axis. The column has a length of 3 m, and the diameter tapers from 1.4 m at the base to 0.7 m at the top. The central hole has a diameter of 0.4 m. Calculate the height of the mass centre above the base. The volume of a complete cone is ($\frac{1}{3}$ base area) $\times$ height, and the mass centre of a uniform solid cone is one-quarter the height above the base.
Ans. 1.1263 m.
8. A parabola has the equation:

$$y = 1.5x^2 + 2$$

Calculate the area under the curve from $x = 2$ to $x = 4$ and calculate the distance of the centroid of this area from the y-axis.
Ans. 32 units; 3.1875 units.

[illegible] has the form of a truncated cone with a
cylindrical hole along its axis. The cylinder has a length [illegible] and
[illegible]
[illegible] Calculate the [illegible]
[illegible] complete cone [illegible]
[illegible] and the mass centre of a uniform solid cone is [illegible]
the [illegible] above its base.

[illegible]

[illegible]

Calculate [illegible]
the [illegible]

[illegible]

Programme 2: Elementary Kinematics

1

Kinematics is concerned with the *motion* of bodies. It is not to be confused with Kinetics, which is the study of the motion of bodies in relation to the forces causing and affecting the motion: this is dealt with in programme 3. Kinematics deals only with the motion itself, and establishes certain rules and formulae for dealing with the problems associated with motion.

The concept of motion involves *displacement*—in fact, displacement *is* motion. If we state that a body is displaced 4 metres from a given point, we are saying that it has moved that distance from the point. Displacement may be measured in any convenient unit of length: metres, light-years, microns, miles, kilometres, are five examples. But for many years, scientific and engineering bodies have adopted the S.I. system of units. (S.I. stands for 'Système International'). In this system, the three fundamental quantities that will concern us—Length, Time and Mass—are always measured respectively in metres (abbreviated to 'm'), seconds (abbreviated to 's') and kilograms (abbreviated to 'kg'). (There are four other fundamental quantities which need not concern us.) You must, therefore, accustom yourself to working with the fundamental unit of the *metre*. Originally (in 1795) the metre was defined by the National Assembly of France to be one ten-millionth part of the distance along a meridian from the north pole to the equator. Metal 'standards' were made, as accurately as the conditions of the time permitted. It was eventually recognised to be impractical to refer back to the original definition, and in 1872 the decision was taken that the metal standard itself should be the definition of the metre. The standard metre was thus defined as the distance between two engraved lines on a platinum–iridium bar which was located at the International Bureau of Weights and Measures, at Sèvres, in the suburbs of Paris. Most developed countries accepted this definition, and retained their own standards, which were accurate copies of the Paris standard, and which were periodically returned to Paris for comparison with the original. This definition of the metre is thus seen to be arbitrary, in that it was determined only by the distance between two marks on a bar.

In 1960 this arbitrary definition of the standard metre was abandoned, and at the present time the metre is defined in terms of the wavelength of krypton-86. It is still, however, an arbitrary definition in that it does not depend upon any other scale of measurement (such as time or mass). In Dimension theory, length is stated to be a fundamental dimension.

Displacement is a *vector*. A vector quantity is one which requires both magnitude and direction to define it completely. To state that an aircraft has travelled 100 kilometres from its base does not tell us where it is: we also require to know the direction in which it went. In this program, however, the vector nature of displacement need not concern us.

2

Velocity is the rate of change of displacement—the change of displacement in unit time. Again, various units are employed. Kilometres per hour, miles per hour, metres per second, knots, are all common and accepted units of velocity. But since we are to be committed to the use of the S.I. system of units, and velocity is defined in terms of length and time, the correct units for velocity must be metres per second, usually abbreviated to m/s, or m s^{-1}. Velocity stated in any other units must first be converted to metres per second before being used in calculations. When a body travels in a straight line with a constant velocity v, it is easy to see that the distance x travelled in time t is given by

$$x = v \times t$$

and in using this simple formula, you have to take care that your units are consistent. If an aircraft travels on a straight course at 600 miles per hour for four hours, we can see, using this formula, that the distance travelled is 2400 miles, and it would be pedantic to insist on converting the speed to metres per second and the time to seconds in order to solve such a trivial problem. But we do not study dynamics to solve trivial problems, and as a general rule, you should always convert all data to the basic units of the S.I. system. The examples in the following frame will afford some practice for you.

Velocity is also a vector. To state that a ship is 450 miles south-east of New York and travelling at 18 knots does not fully define its motion; the additional information that it is travelling due east completes the picture. When we are concerned with motion in a straight line only, the vector nature of velocity need not concern us, except to the extent that we must be aware that a body travelling along a straight path may travel either way, warning us that we must be alive to the necessity of using positive and negative signs. Looking further ahead to Frame 51, when motion along a circular path is examined, we shall see that the vector nature of velocity becomes important.

We referred in Frame 1 to Dimension theory. This is a convenient point to expand a little on this topic. A 'Dimension' is best thought of as any quantity which must be measured, as distinct from being calculated. For example, once we have agreed on a definition of a standard metre, we shall need some device to measure length, such as a metre rule, a micrometer or a surveyor's chain, which directly or indirectly relates to our standard metre. Similarly, having defined one second of time, we need watches, clocks or atomic clocks, to measure it. Length (L) and Time (T) are thus two *fundamental* Dimensions used in mechanics. But when we come to Velocity, we find that no additional measurement is required; we *calculate* velocity from measured length and time. The Dimension of velocity is accordingly said to be *derived*, as distinct from fundamental. It may be consoling to learn that in mechanics, so far as we are concerned, the fundamental dimensions are limited to three: Length (L), Time (T) and Mass (M). All other quantities are derived from these three. For example, the dimensions of velocity would be L/T. Equations of mechanics must always be correct as to dimensions; that is, the dimensions of one side of the equation must be the same as those of the other. Taking the very simple equation above, the dimensions of x, the displacement, must be L. The resulting dimension of the right-hand side are $(L/T) \times T = L$. The equation is thus dimensionally correct.

3

Speed is similar to velocity, and is measured in the same units, but it is not a vector. It is simply the magnitude of velocity. (In mathematical terms, it is the 'modulus' of the velocity vector.) In this programme, this distinction between speed and velocity need not concern us.

You must be capable of converting quantities from any one system of units to any other. Although nearly all textbooks on mechanics present all their calculations in the S.I. system, you cannot expect real problems always to be so conveniently ordered. Conversion of units may appear simple; indeed, it is, but it is one area in which students are prone to make absurd mistakes, usually because students are not noted for critically examining the results of their calculations. A student has been known to write quite happily that a time of $2\frac{1}{2}$ hours is the same as 0.0006944 seconds, and to fail completely to see the absurdity of the statement, until it is pointed out to him.

The following conversion factors should suffice for your needs.

$$1 \text{ hour} = 3600 \text{ seconds} \left(\frac{3600 \text{ s}}{1 \text{ hour}}\right)$$

$$1 \text{ minutes} = 60 \text{ seconds} \left(\frac{60 \text{ s}}{1 \text{ min}}\right)$$

$$1 \text{ mile} = 1609.34 \text{ metres} \left(\frac{1609.34 \text{ m}}{1 \text{ mile}}\right)$$

$$1 \text{ yard} = 0.9144 \text{ metres} \left(\frac{0.9144 \text{ m}}{1 \text{ yard}}\right)$$

$$1 \text{ foot} = 0.3048 \text{ metres} \left(\frac{0.3048 \text{ m}}{1 \text{ ft}}\right)$$

$$1 \text{ inch} = 0.0254 \text{ metres} \left(\frac{0.0254 \text{ m}}{1 \text{ in}}\right)$$

The expressions to the right of each conversion factor are called Unity brackets. They afford a simple and (almost) foolproof way of converting from one system to another. You employ the brackets either as they are, or the other way up, according to what unit you are converting to; by writing the units themselves in addition to the numbers, you so arrange your expression that all the unwanted units cancel out. Here is the calculation for converting 15 miles per hour to metres per second.

$$15 \text{ mile/hour} = 15\,\frac{\cancel{\text{mile}}}{\cancel{\text{hour}}} \times \left(\frac{1\,\cancel{\text{hour}}}{3600\text{ s}}\right) \times \left(\frac{1609.34\text{ m}}{1\,\cancel{\text{mile}}}\right)$$

$$= \underline{6.7056\text{ m/s}}$$

Notice how the first unity bracket has been inverted, so that 'hour' cancels 'hour' on the bottom line of 15 miles per hour, and how 'mile' is cancelled by 'mile' on the bottom line of the second unity bracket.

For practice, do the following conversions yourself. The answers (to four places of decimals, where necessary) are given in brackets.

Convert

20 mile/hour to m/s	(8.9408 m/s)
60 ft/s to m/s	(18.288 m/s)
45 kilometres/hour to m/s	(12.5 m/s)
8 metres/minute to inches/s	(5.2493 in/s)
840 mm/min to m/s	(0.014 m/s)
840 mm/min to mile/hour	(0.0314 mile/hour)
60 mile/hour to feet per second	(88 ft/s)

4

Velocity is not always constant. Although the average velocity during a car journey might be 28 miles per hour, the actual velocity at any time might be anything from 70 miles per hour to zero. (It could even be negative, if the driver had forgotten something, and had to go back!) The rate of change of velocity is called *acceleration*. Negative acceleration (that is, slowing down) is called *retardation*, or sometimes, rather inelegantly, deceleration. Since velocity is a vector, it follows that acceleration also must be a vector, but again, this need not concern us in this program.

Acceleration is the rate of change of velocity, or the change of velocity divided by the time taken. Its units must therefore be

(metres per second) per second

This is usually written as m/s^2, or, preferably, m s^{-2}.

Here is a little problem. A car starts from rest. It increases speed at a steady rate and attains 54 kilometres per hour (km/h) in 6 seconds. What is its acceleration? Have a try at working it out. The solution follows in the next frame.

5

$$54 \text{ km/h} = 54 \times 1000 \text{ m/h} = \frac{54000}{60 \times 60} \text{ m/s} = 15 \text{ m/s}$$

The acceleration, a, is therefore given by

$$a = \frac{15 \text{ metres per second}}{6 \text{ seconds}} = \underline{2.5 \text{ m s}^{-2}}$$

But of course, all problems cannot be solved so simply from first principles.

Consider this next problem.

A car is travelling at 8 m/s and after 3 seconds its speed is 14 m/s. What distance does it cover in the 3 seconds?

You should be able to work out in your head that the acceleration is 2 m s^{-2} (a speed increase of 6 m/s in 3 s) but the question unfortunately does not ask this. See if you can work out the required distance before consulting Frame 6.

6

You may have reasoned something along the following lines

"The car starts at 8 m/s and increases to 14 m/s. So its average speed will be 11 m/s. So in the three seconds, it will travel $(11 \times 3) = 33$ m"

This is a reasonable approach to the solution of the problem, but it begs the question: "What do we mean by 'average'?" To save you some anxiety at this point, it is only fair to state that the problem *as stated* cannot be solved. Information should have been given about *how* the speed increased. For example, the question could have been put

"A car increases its speed *uniformly* from 8 m/s to 14 m/s is 3 s . . ." etc.

and with this information, the argument about average speed is valid. Let us look at two sketches of graphs of velocity against time, the first graph showing a uniform rate of increase, and the second a non-uniform increase.

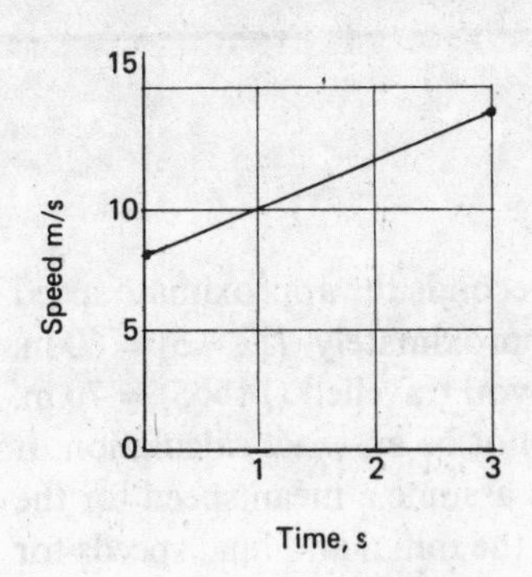

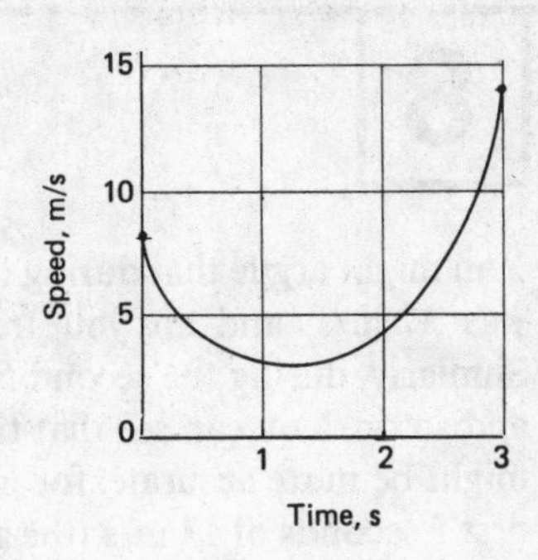

The first graph shows that the average speed will be $\frac{1}{2}(8+14) = 11$ m/s. We cannot calculate the average speed for the second case, because we do not know exactly how the speed varies. We could say approximately that the speeds at 0, 1, 2 and 3 seconds are 8, 3, 4 and 14 m/s, giving an approximate average of $\frac{1}{4}(8+3+4+14) = 7\frac{1}{4}$ m/s, and that consequently the distance covered would be approximately $(7\frac{1}{4} \times 3) = 21\frac{3}{4}$ m. We shall look more closely at the problem of non-uniform change of speed in the next frame.

7

Now although we shall not concern ourselves in this programme with problems involving non-uniform speed change (or 'variable acceleration', to use the more correct terminology) we shall find it valuable to consider such a state of affairs. Imagine yourself as a co-driver in a car rally. While your partner is driving, you are recording the speed of the car at regular intervals—say, every 5 seconds. After 60 seconds, your record may look like this:

Time (s)	0	5	10	15	20	25	30	35	40	45	50	55	60
Speed (m/s)	12	14	16	17	17	15	14	12	11	8	5	7	10

How would you make use of this record to estimate how far you have travelled during the 60 seconds? See if you can formulate a procedure before reading on.

8

You might argue that during the first 5 seconds the approximate speed was 12 m/s, and so you travelled approximately $(12 \times 5) = 60$ m. Similarly, during the second 5 seconds, you travelled $(14 \times 5) = 70$ m, and so on. You can see that this would not be an exact calculation. It might be more accurate, for instance, to assume a mean speed for the first 5 seconds of 13 m/s (the average of the initial and final speeds for the period of 5 seconds). But either calculation would give you a fairly accurate estimate of the total distance covered. Using the first method, the total distance works out at 740 m and the second method gives 735 m—not a big difference. You can check these figures yourself.

Here is a velocity–time graph for this journey.

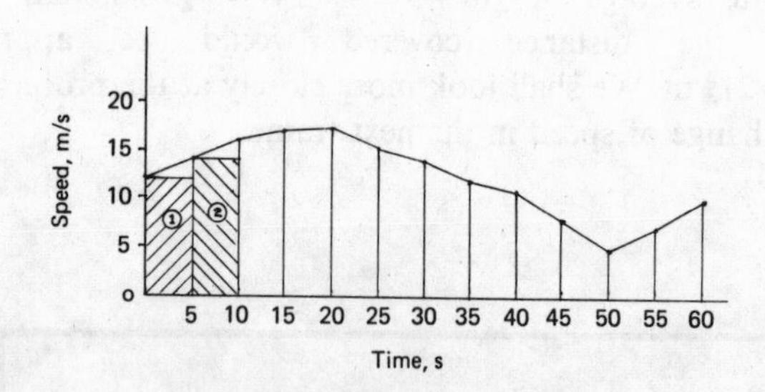

You can see that the first calculation of (12×5) m is the area of the first column of the graph, shaded, and marked 1 . (14×5) is the area of the second column, and so on. The little areas between the tops of the columns and the curve are errors arising out of the assumption that 12 m/s is the speed for the whole 5 seconds. Imagine you were able to record speed every second instead of every 5 seconds. The first 5-second period would then look like

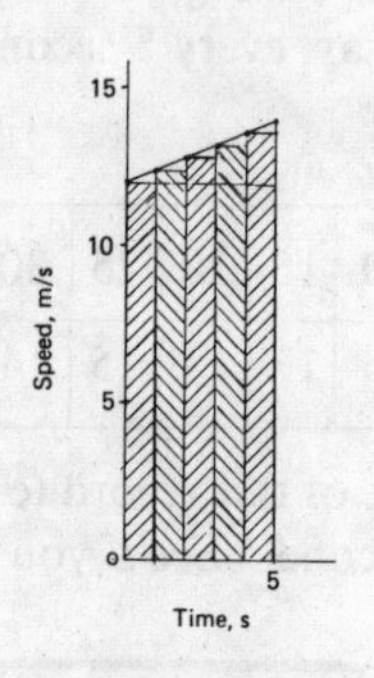

and you can see that the 'errors'—the bits between the tops of the columns and the curve—are now much less over the 5 seconds than they were. So if you could obtain a *continuous* reading of velocity with time, and draw the graph, the *area under the graph would give you the exact distance travelled.*

9

The conclusion reached in Frame 8 is very important, and you will almost always find that the sketching of a velocity–time graph will be a great help in solving problems of linear kinematics. If you look back to the first of the two graphs of Frame 6, the area under this graph will be Base × Mean height $= 3 \times \frac{1}{2}(8 + 14) = 33$ m, which is the answer required.

We can use this knowledge to derive some useful equations relating velocity, time, displacement and acceleration. We can begin by drawing a velocity–time graph for a body which starts off with a velocity u, and increases *uniformly* to a velocity v in a time t.

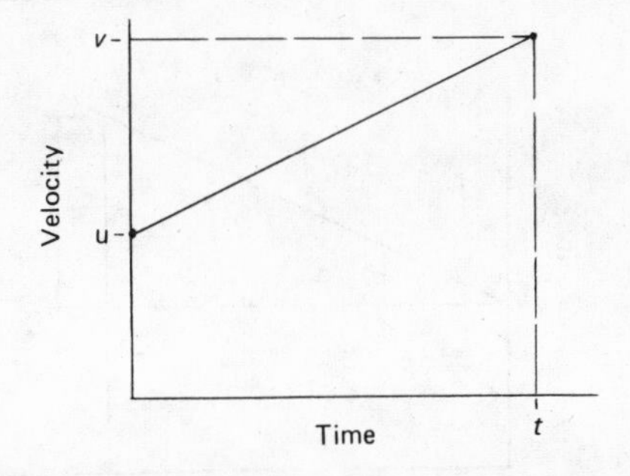

Derive a formula for the velocity v after time t, in terms of u, a and t. Remember that a is the increase of velocity in unit time.

10

a is the increase of velocity in unit time. So the increase in time $t = (at)$. So the final velocity will be $u + (at)$. So our formula is

$$v = u + at \qquad (1)$$

Now make use of the principle stated in Frame 8 to derive a second formula.

11

Distance = area under velocity–time graph. Therefore

$$x = t \times \tfrac{1}{2}(u + v)$$

or

$$x = \tfrac{1}{2}t(u + v) \qquad (2)$$

There is another way we can reckon the area under the graph.

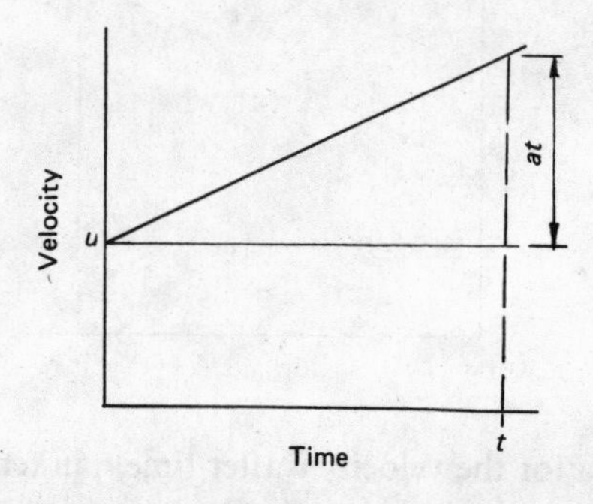

The increase of velocity in time t is (at) as we found in Frame 10. So we can divide the area into a rectangle and a triangle. Express the area as the sum of these two to get our third formula.

12

The area of the rectangle is $(u \times t)$.
The area of the triangle is $\frac{1}{2} \times t \times (at)$.
Adding the two

$$x = ut + \tfrac{1}{2} \times t \times (at)$$

or

$$x = ut + \tfrac{1}{2}at^2 \qquad (3)$$

A fourth formula can be found by taking any two of the ones we have got, and eliminating t from them. Do this. One solution is given in Frame 13, but the same formula can be obtained using *any* two of the three formulae.

13

Taking formulae (1) and (2)

$$v = u + at$$
$$x = \tfrac{1}{2}t(u+v)$$
$$\therefore \; t = \frac{v-u}{a} = \frac{2x}{u+v}$$
$$\therefore \; 2ax = (v-u)(v+u) = v^2 - u^2$$
$$\therefore \; v^2 = u^2 + 2ax \qquad (4)$$

For convenience the four formulae are reproduced below:

$$v = u + at \qquad (1)$$
$$x = \tfrac{1}{2}t(u+v) \qquad (2)$$
$$x = ut + \tfrac{1}{2}at^2 \qquad (3)$$
$$v^2 = u^2 + 2ax \qquad (4)$$

It is probable that you have seen these formulae before, although you may have used a different notation. Learners are frequently puzzled as to which formula to use; there seem to be so many. If we set them up in the form of a table

	u	v	x	a	t
No. 1	/	/		/	/
No. 2	/	/	/		/
No. 3	/		/	/	/
No. 4	/	/	/	/	

you can now see that each formula connects a different four of the five variables. (You could obtain a fifth formula by eliminating u from any two of the above, but four are enough for us.) As to which formula to use to solve any problem, the rule is: list all the quantities given, together with the quantity required, and use the formula which relates all these.

If you have done calculus, you can derive the formulae starting with

$$a = \frac{dv}{dt}$$

and integrating. This is done in Frame 14. If you have not yet learned to use this powerful method, it does not matter. Just skip Frame 14 and go straight on to Frame 15.

14

$$a = \frac{dv}{dt}$$

$$\therefore \int dv = \int a\, dt$$

$$\therefore v = at + C_1$$

Writing 'initial condition': when $t = 0$, $v =$ initial velocity u. Therefore

$$u = a \times 0 + C_1; \quad \therefore C_1 = u$$

$$\therefore v = u + at \tag{1}$$

$$v = \frac{dx}{dt}$$

$$\therefore \frac{dx}{dt} = u + at$$

$$\therefore \int dx = \int (u + at)\, dt$$

$$\therefore x = ut + \tfrac{1}{2}at^2 + C_2$$

When $t = 0$, $x = x_0$. But this is always assumed to be 0.

$$\therefore 0 = 0 + 0 + C_2; \quad \therefore C_2 = 0$$

$$\therefore x = ut + \tfrac{1}{2}at^2 \tag{3}$$

Rewriting $a = \frac{dv}{dt}$ as 'function of a function'

$$a = \frac{dv}{dt} = \frac{dv}{dx} \times \frac{dx}{dt} = \frac{dv}{dx} \times v$$

$$\therefore \int a\, dx = \int v\, dv$$

$$\therefore ax = \tfrac{1}{2}v^2 + C_3$$

When $x = x_0 = 0$, $v = u$

$$\therefore 0 = \tfrac{1}{2}u^2 + C_3; \quad C_3 = -\tfrac{1}{2}u^2$$

$$\therefore ax = \tfrac{1}{2}v^2 - \tfrac{1}{2}u^2$$

$$\therefore v^2 = u^2 + 2ax \tag{4}$$

From (1) and (3)

$$a = \frac{v-u}{t}$$

$$\therefore\ x = ut + \tfrac{1}{2}t\left(\frac{v-u}{t}\right)$$

$$\therefore\ x = \tfrac{1}{2}(u+v)t \qquad (2)$$

15

Now that we have the four formulae, let us see how they are used to solve problems of motion. Look at this problem.

Problem A car travels a distance of 48 m in 8 seconds with an initial velocity of 12 m/s. Calculate the acceleration.

Select the correct formula, using the rule given in Frame 13. This was: list the quantities given, and the quantity required, and find the formula which includes all these.

16

We are given the initial velocity u (12 m/s), the displacement x (48 m) and the time t (8 s). We are asked for the acceleration a. The formula relating these four variables is Formula (3)

$$x = ut + \tfrac{1}{2}at^2$$

Substituting

$$48 = 12 \times 8 + \tfrac{1}{2} \times a \times 8^2$$

$$\therefore\ a = \frac{48-96}{\frac{1}{2}\times 64} = \underline{-1.5\ \text{m s}^{-2}}$$

The acceleration is negative; the car is therefore slowing down. This becomes obvious when we realise that if it travelled for 8 seconds at the initial speed of 12 m/s it would cover a distance of $(8 \times 12) = 96$ m. Since it travels only 48 m, the average speed must be less than the initial value of 12 m/s.

Now calculate the final velocity. This is another problem, requiring a different formula.

17

You could use Formula (1)

$$v = u + at = 12 + (-1.5)8 = \underline{0\ \text{m/s}}$$

Or you may use Formula (4):

$$v^2 = u^2 + 2ax = 12^2 + 2(-1.5)48 = 144 - 144 = 0\ \text{m/s}$$
$$\therefore\ v = \underline{0\ \text{m/s}}$$

Or you may use Formula (2)

$$x = \tfrac{1}{2}t(u + v)$$
$$48 = \tfrac{1}{2} \times 8(12 + v)$$
$$\therefore\ v = \frac{96}{8} - 12 = \underline{0\ \text{m/s}}$$

The last solution is the 'correct' one because the other two make use of the value of a calculated before, and if we had made a mistake in the calculation of a the subsequent calculations would also have been wrong.

18

Here is another problem for you to try—or rather, three problems in one.

Problem A car with constant acceleration passes a point at 8 m/s and then passes a second point 1 km distant at 20 m/s. Determine (a) the average speed; (b) the time taken; (c) the acceleration.

Part (a) you can do in your head; no formula is needed. Select the appropriate formula to solve (b). Then solve part (c) taking care, preferably, to avoid using the result of part (b). You may, of course, use this if you wish, but there is always the risk that your calculated 'time taken' might be wrong. The next frame contains the solution for you to check your attempt against.

19

(a) The average velocity is, simply: $\frac{1}{2}(8+20) = \underline{14 \text{ m/s}}$.

(b) Given: u (8 m/s), v (20 m/s), x (1000 m). Required: t.
Use Formula (2)

$$x = \tfrac{1}{2}t(u+v)$$

$$\therefore\ 1000 = \tfrac{1}{2}t(8+20)$$

$$\therefore\ t = \frac{2000}{28} = \underline{71.43 \text{ s}}$$

(c) Given: u, v, x. Required: a.
Use Formula (4)

$$v^2 = u^2 + 2ax$$

$$20^2 = 8^2 + 2a \times 1000$$

$$\therefore\ a = \frac{400-64}{2000} = \frac{336}{2000} = \underline{0.168 \text{ m s}^{-2}}$$

20

Sometimes motion occurs in two or more stages.

> *Problem* A vehicle starts from rest and reaches a maximum speed $\hat{v}$ with a constant acceleration of 1.25 m s^{-2}. It then comes to rest with a uniform retardation of 1.875 m s^{-2}. The total distance travelled is 150 m. Find the maximum velocity $\hat{v}$ and the total time taken.

There are no general rules for the solution of this type of problem but it is always a good plan to make a sketch of the velocity–time graph. You may also learn something by sketching an acceleration–time graph. Call the time to reach maximum speed t_1 and the time to return to rest t_2. Total distance covered will be the area under the velocity–time graph. Use Formula (1) twice—to relate $\hat{v}$ and t_1, and to relate $\hat{v}$ and t_2. Frame 21 gives the complete solution.

21

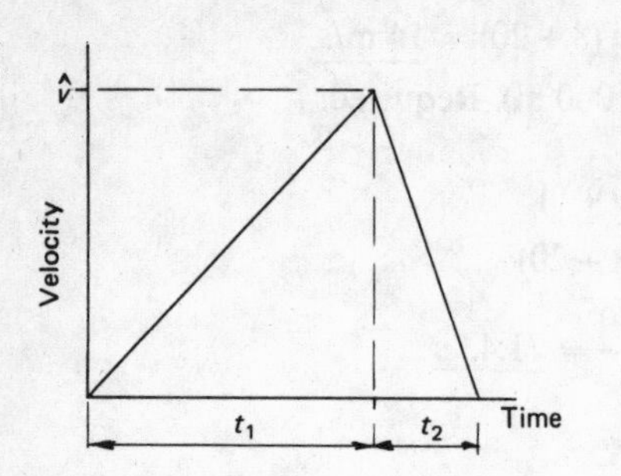

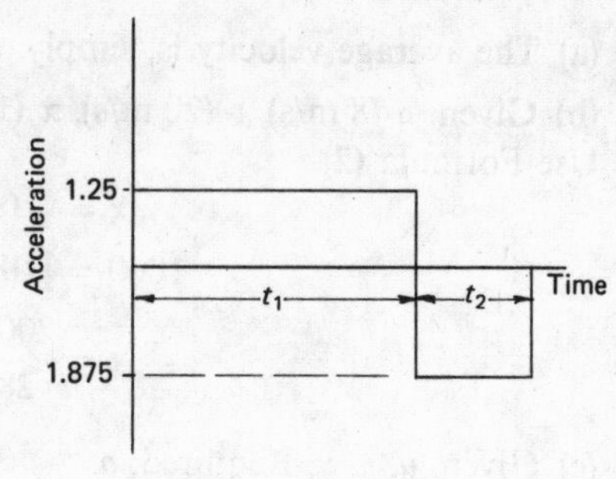

The acceleration during the time t_1 has a constant value of 1.25 m s^{-2}; this part of the acceleration–time graph is thus the horizontal straight line, and because the velocity increases, this acceleration is positive. During the second period t_2 the acceleration has a constant negative value of 1.875 m s^{-2}. This explains the form of the acceleration–time graph. Also, acceleration is the time-derivative of velocity; thus velocity is the time-integral of acceleration. So that the *area* under an acceleration–time graph will give the change of velocity in that time, in the same manner that the area under a velocity–time graph gives the displacement in that time.

Displacement = area under velocity–time graph

$$150 = \tfrac{1}{2}(t_1 + t_2)\hat{v}$$

$$\therefore \hat{v}(t_1 + t_2) = 300 \qquad (1)$$

Applying Formula (1) to both stages

$$\hat{v} = 0 + 1.25t_1; \qquad 0 = \hat{v} - 1.875\,t_2$$

$$\therefore t_1 = \frac{v}{1.25}; \qquad \therefore t_2 = \frac{v}{1.875}$$

Substituting these values in equation (1)

$$\hat{v}\left(\frac{\hat{v}}{1.25} + \frac{\hat{v}}{1.875}\right) = 300$$

$$\therefore v^2(0.8 + 0.533) = 300$$

$$\therefore \hat{v} = \sqrt{225} = \underline{15 \text{ m/s}}$$

Substituting

$$t_1 = \frac{\hat{v}}{1.25} = \frac{15}{1.25} = \underline{12\ \text{s}}$$

$$t_2 = \frac{\hat{v}}{1.875} = \frac{15}{1.875} = \underline{8\ \text{s}}$$

$$\text{Total time} = \underline{20\ \text{s}}$$

22

The next problem is slightly harder, but you can solve it using practically the identical procedure to that just used, that is, obtain one equation from the area under the graph, and two more by making use of Formula (1) on both stages of the motion.

Problem A vehicle accelerates uniformly from rest at 2 m s^{-2} to a velocity $\hat{v}$. It then accelerates at 1.4 m s^{-2} to a final velocity of 34 m/s. The total distance travelled is 364 m. Calculate the value of $\hat{v}$ and the times taken for the two stages of the journey.

The required answers are 16 m/s, 8 s and 12 s. The solution follows.

23

Here is the complete solution.

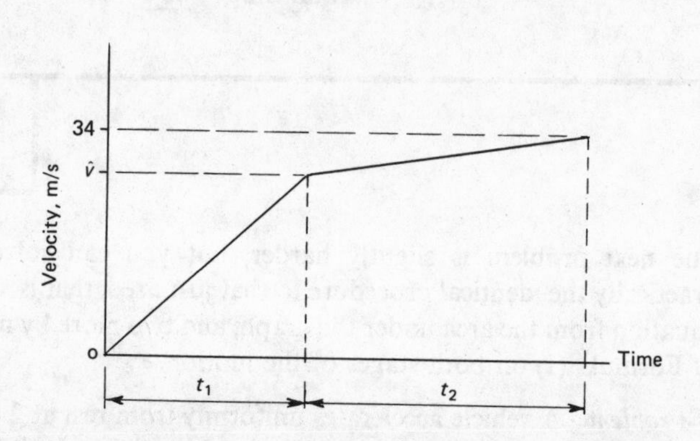

Equating the area under the graph to the total distance travelled

$$\tfrac{1}{2}\hat{v}t_1 + \tfrac{1}{2}(\hat{v}+34)t_2 = 364$$

Using Formula (1)

$$\hat{v} = 0 + 2t_1; \qquad 34 = \hat{v} + 1.5t_2$$

Substituting for t_1 and t_2

$$\tfrac{1}{2}\hat{v}\left(\frac{\hat{v}}{2}\right) + \tfrac{1}{2}(\hat{v}+34)\left(\frac{34-\hat{v}}{1.5}\right) = 364$$

Multiplying by 12

$$3\hat{v}^2 + 4(34^2 - \hat{v}^2) = 4368$$

$$3\hat{v}^2 + 4624 - 4\hat{v}^2 = 4368$$

$$\therefore\ \hat{v} = \sqrt{4624 - 4368} = \underline{16 \text{ m/s}}$$

$$\hat{v} = 2t_1; \qquad \therefore\ t_1 = 8 \text{ s}$$

$$34 = \hat{v} + 1.5t_2; \qquad \therefore\ t_2 = \frac{34-16}{1.5} = \underline{12 \text{ s}}$$

An alternative solution would be to use Formula (4) on both stages of the motion, calculating the displacements x_1 and x_2 for the two stages. The sum of these displacements is 364 m. If you think you need practice, do this yourself, but a solution is not given.

24

Here is one last problem before we go on to the 'drill' exercises. No hints at all are now given. See how you get on with it.

Problem A car starts from rest and reaches a maximum velocity $\hat{v}$ with a constant acceleration of 2 m s^{-2}. It travels at this maximum speed for a time, and then retards at a constant rate of 3 m s^{-2} to rest. The total distance travelled is 945 m and the total time taken is 60 s. Calculate the maximum velocity $\hat{v}$ and the times for the three stages of the journey.

A complete solution follows. But do not forget that it is not the only possible solution.

25

Always sketch the velocity–time graph

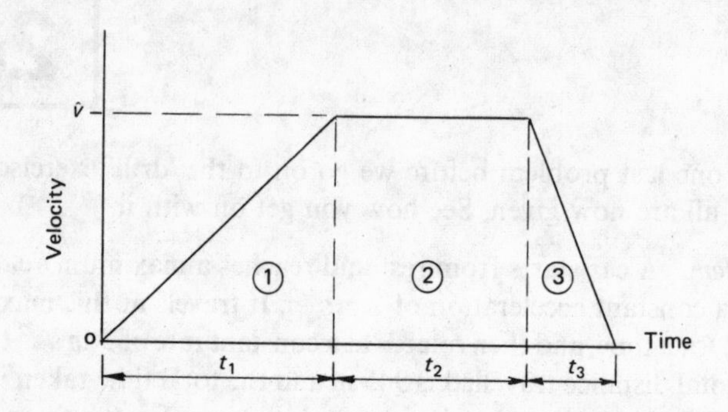

Use Formula (2) three times

For stage 1: $x_1 = \frac{1}{2}(0+\hat{v})t_1$; $\therefore\ x_1 = \frac{1}{2}\hat{v}t_1$

For stage 2: $x_2 = \frac{1}{2}(\hat{v}+\hat{v})t_2$; $\therefore\ x_2 = \hat{v}t_2$

For stage 3: $x_3 = \frac{1}{2}(\hat{v}+0)t_3$; $\therefore\ x_3 = \frac{1}{2}\hat{v}t_3$

Adding these

$$945 = \tfrac{1}{2}\hat{v}t_1 + \hat{v}t_2 + \tfrac{1}{2}\hat{v}t_3 \qquad \text{(a)}$$

Using Formula (1)

For stage 1: $\hat{v} = 0 + 2t_1$; $\therefore\ t_1 = \dfrac{\hat{v}}{2}$

(b)

For stage 3: $0 = \hat{v} - 3t_3$; $\therefore\ t_3 = \dfrac{\hat{v}}{3}$

Substitute these values in equation (a)

$$945 = \frac{\hat{v}^2}{4} + \hat{v}t_2 + \frac{\hat{v}^2}{6} = \frac{5\hat{v}^2}{12} + \hat{v}(60 - t_1 - t_3)$$

$$945 = \frac{5\hat{v}^2}{12} + \hat{v}\left(60 - \frac{\hat{v}}{2} - \frac{\hat{v}}{3}\right)$$

Multiply by 12

$$11340 = 5\hat{v}^2 + 720\hat{v} - 10\hat{v}^2 = 720\hat{v} - 5\hat{v}^2$$

Re-arranging

$$\hat{v}^2 - 144\hat{v} + 2268 = 0$$

$$\begin{aligned}\therefore\ \hat{v} &= 72 \pm \sqrt{5184 - 2268}\\ &= 72 \pm 54\\ &= \underline{18\ \text{m/s}}\ \text{or}\ 126\ \text{m/s}\end{aligned}$$

Substituting in eqns (b)

$$t_1 = \frac{18}{2} = \underline{9\ \text{s}}; \qquad t_3 = \frac{18}{3} = \underline{6\ \text{s}}$$

$$\therefore\ t_2 = 60 - (9 + 6) = \underline{45\ \text{s}}$$

This completes the solution, but some remarks follow in Frame 26.

26

Two points arise out of this problem. The first concerns the determination of $\hat{v}$. We obtain this by solving a quadratic equation. You will have seen that we took the value of 18 m/s as the correct value, and rejected 126 m/s. Why? Well, a value of 126 would certainly be a legitimate solution so far as the algebra were concerned, but when we evaluate t_1 and t_2, we should find that we obtained 63 s and 42 s respectively, leaving a value of t_2 of -45 s. It is clearly impossible that the car should travel at this speed over the second stage of the journey for a *negative* time, and so, we reject this part of the solution. Nevertheless, one should always investigate the two answers to the solution of a quadratic equation, as frequently they may have a real physical interpretation, and, indeed, be both essential solutions to the problem. This is particularly true in the case of freely falling bodies, which we shall look at next. In this particular problem, the second value of $\hat{v}$ of 126 m/s does, oddly enough, have a physical significance. Although there can be no such thing as 'negative time' in this context, you will find out, if you care to check, that the conditions of the problem will be met if the car travels for 63 s with acceleration 2 m s^{-2}; then travels *in the opposite direction* at 126 m/s for 45 s, and finally travels forwards again for 42 s, retarding at 3 m s^{-2}.

The second point arising is that, with rather involved problems of this sort, it is good practice to check your answers. In this case, let us check that with the values of $\hat{v}$, t_1, t_2 and t_3 calculated, the actual distance covered will be correct.

$$x_1 = \tfrac{1}{2}\hat{v}t_1 = \tfrac{1}{2} \times 18 \times 9 = 81 \text{ m}$$
$$x_2 = \hat{v}t_2 = 18 \times 45 = 810 \text{ m}$$
$$x_3 = \tfrac{1}{2}\hat{v}t_3 = \tfrac{1}{2} \times 18 \times 6 = 54 \text{ m}$$

Adding

$$\text{Total distance} = \underline{945 \text{ m}}$$

You now need practice, and Frame 27 contains ten exercises for you to work on. Do not be too discouraged if you find that they take you a long time, or even if there are some that you cannot do, and need to ask for help. And remember that there may be two or three ways of solving a problem: all of them may be 'right' but some ways will be easier than others. It is also fair to tell you that although this is an 'elementary' text, you should not assume that the problems at this stage are necessarily going to be easy. You may even find that some of your lecturers cannot do some of them straight away, without a bit of preparation.

27

'Drill' exercises: motion in a straight line

1. A vehicle moves at 5 m/s with an acceleration of 1.2 m s^{-2}. Calculate its velocity after 10 s, and the distance travelled.
   Ans. 17 m/s; 110 m.
2. A train can retard at a maximum rate of 2 m s^{-2}. Find the shortest distance and time to bring it to rest from a speed of 95 km/hour.
   Ans. 174.1 m; 13.19 s.
3. A body with initial speed 5 m/s and uniform acceleration moves 600 m in 70 s. Calculate the acceleration, and the maximum speed.
   Ans. 0.102 m s^{-2}; 12.14 m/s.
4. Two signals on a railway are 950 m apart. A train passes the first one at a speed of 96 km/hour with uniform retardation, and it passes the second one after 40 s. Find the retardation, and the final velocity, in km/hour.
   Ans. −0.1458 m s^{-2}; 75 km/hour.
5. A car travels with constant retardation, and passes a point at 16 m/s, and a second point 500 m distant, at 4 m/s. Find the average speed, the time taken, and the retardation.
   Ans. 10 m/s; 50 s; 0.24 m s^{-2}.
6. The cage of a mine hoist descends part of the shaft with an acceleration of 1 m s^{-2} and the remainder of the distance with a retardation of 2 m s^{-2}. The total time taken for the descent is 60 s. Find the shaft depth and the maximum speed.
   Ans. 1200 m; 40 m/s.
7. Three points on a track are spaced at 50 m intervals. A car with uniform acceleration passes them, taking 8 s to travel from the first to the second, and 7 s from second to third. Determine the acceleration, and the speeds at each of the three points.
   Ans. 0.119 m s^{-2}; 5.77 m/s; 6.73 m/s; 7.56 m/s.
8. The cage of a hoist takes 9 s to descend the shaft. For the first quarter of the distance, it accelerates uniformly, and for the last quarter, it retards uniformly, the middle section being covered at uniform speed. The shaft depth is 40 m. The rate of retardation is the same as the acceleration. Determine this acceleration, and the maximum speed.
   Ans. 2.22 m s^{-2}; 6.667 m/s.

9. An elevator can accelerate at $2\ \mathrm{m\,s^{-2}}$ and can retard at $6\ \mathrm{m\,s^{-2}}$. Find the least time for it to ascend 40 m (a) if the maximum speed is 4 m/s, (b) if the maximum speed is unlimited. In the latter case, what will be the maximum speed?
   Ans. 11.33 s; 7.3 s; 10.95 m/s.
10. A vehicle travels 4.8 km in 5 minutes. For the first stage, it accelerates at $1.6\ \mathrm{m\,s^{-2}}$ from rest, and reaches a maximum velocity $\hat{v}$, at which speed it covers the second stage. For the third stage, it retards uniformly at $2.5\ \mathrm{m\,s^{-2}}$ to rest. Calculate the times for each stage, and the maximum velocity.
    Ans. 10.28 s; 283.1 s; 6.58 s; 16.45 m/s.

28

The equations derived for the solution of problems of straight-line motion can be used to analyse the motion of bodies subjected to the force of gravity. Objects close to the surface of the earth, and subjected only to their own weight, are observed to have an acceleration which is approximately constant, at all points on the earth, and which is independent of the mass of the body itself. This acceleration is approximately 9.81 $m\,s^{-2}$, and it is always directed vertically downwards, whether the body is moving upwards or downwards. Thus, such a body moving upwards will slow down in speed, while a body moving downwards will increase in speed. The reason for this phenomenon will become clearer when we look at Elementary Kinetics. The value of this acceleration does actually vary a small amount; it is, for example, slightly less at the equator than it is at the poles, but this difference is very small, and for most engineering claculations (and for all calculations in this programme) the figure of 9.81 $m\,s^{-2}$ is considered to be sufficiently accurate. The symbol g is used for this particular value of acceleration.

You must never forget that this 'constant' acceleration is really only an approximation; the true value of acceleration varies inversely as the distance of the body from the earth's centre. Thus, it would be wrong to assume that the acceleration of a satellite such as Telstar was 9.81 $m\,s^{-2}$ because it is at a considerable height above the earth's surface—several thousand kilometres, in fact. At this distance, the acceleration due to gravity is approximately only 0.25 $m\,s^{-2}$. In all problems that we consider in this programme, however, we may assume our bodies to be close enough to the earth's surface to justify assuming a constant value for g of 9.81 $m\,s^{-2}$.

You should note particularly in the first paragraph of this frame the words 'subjected only to their own weight'. Now objects falling or rising close to the earth's surface are not subject only to their own weight; there will be in addition some resistance due to the motion through the air. The effect of air resistance on the motion of a stone falling from a building is likely to be very slight, and may justifiably be neglected in most cases. But on the other hand, to neglect the effect of air resistance on a man descending from a plane by parachute would be absurd. Even without a parachute, the air would have a very pronounced effect on the motion of his body. Without air resistance, the constant acceleration would cause the speed of fall of a body to increase continuously until the body reached the earth. But air resistance increases as the speed of fall increases, and eventually (assuming a body has plenty of space in which to fall) a falling body attains a speed at which the air resistance exactly equals its weight, and the acceleration then is zero. The body now travels at this constant speed, which is called the *terminal velocity*. With a parachute, the terminal velocity is (fortunately) quite low; without a parachute, for a human body, the terminal velocity is approximately 120 miles per hour; for a brick, or a bomb, it would, of course be a good deal higher than this. In all problems in this programme, we shall ignore or neglect the effect of air resistance on the motions of bodies, but ignoring or neglecting something is no excuse for forgetting its existence.

For bodies in free 'fall' (in which is included bodies moving upwards as well) we must always remember to use a sign convention when using our four formulae.

29

Problem A missile is projected vertically upwards with an initial velocity of 80 m/s. Neglecting air resistance, calculate the time for it to reach a height of 200 m and calculate the velocity at this point.

List the quantities given, and the quantity required, not forgetting that the acceleration will now be 9.81 $\mathrm{m\,s^{-2}}$. Select the correct formula and substitute the data. Remember that there are two problems here; just concentrate first on solving the time. As we forecast at the end of the previous frame, you will need to adopt a sign convention, and you should therefore call all upward-directed quantities positive, and downward-directed ones negative. (So what will your acceleration be?)

30

We have been given the initial velocity u (80 m/s), the distance x (200 m) and we assume the acceleration a (9.81 ms^{-2}). We are asked to find the time t.

The correct formula will therefore be Formula (3): $x = ut + \frac{1}{2}at^2$. Having regard to signs

$$u = +80 \text{ m/s}$$
$$x = +200 \text{ m}$$
$$a = -9.81 \text{ m s}^{-2}$$

Substituting

$$200 = 80t + \tfrac{1}{2}(-9.81)t^2$$
$$\therefore\ 200 = 80t - 4.905t^2$$

Simplifying

$$t^2 - 16.31t + 40.77 = 0$$
$$\therefore\ t = 8.155 \pm \sqrt{66.5 - 40.77} = 8.155 \pm 5.072$$

The *two* solutions are $t_1 = \underline{3.083 \text{ s}}$ and $t_2 = \underline{13.227 \text{ s.}}$

It should be clear to you at this stage why there are two answers for t, but if it isn't, proceed with the second part of the problem, which is to calculate the velocity. Choose the 'correct' equation—that is, an equation which does not need to make use of the values of t just calculated: they could be wrong! The solution is completed in the next frame.

31

To determine the velocity, we are again given u, x and a and we require v. Accordingly, Formula (4) is indicated.

$$v^2 = u^2 + 2ax$$
$$= 80^2 + 2(-9.81)200$$
$$= 6400 - 3924$$
$$= 2476$$

$$\therefore\ v = \pm 49.759 \text{ m/s}$$

This gives the *two* solutions,

$v_1 = \underline{+49.759 \text{ m/s}}$ and $v_2 = \underline{-49.759 \text{ m/s}}.$

The first solution tells us that the missile is moving upwards, and the second that it is moving downwards. The explanation for two answers for t is now clear. The time of 3.083 s is the time to reach a height of 200 m on the way up; the time of 13.227 s is the time for the missile to pass the same point on its way back down again.

This example illustrates an important point concerning problems of motion due to gravity. It is not necessary to divide the problem into two parts—one *up* and one *down*—as students are frequently prone to do.

Now use the same data to calculate the maximum height reached by the missile. Choose your formula correctly.

32

Given: u, a and, of course, v, which must be zero at the maximum height. And x, the height, is required. We use Formula (4).

$$v^2 = u^2 + 2ax$$

$$0 = 80^2 + 2(-9.81)x$$

$$\therefore\ x = \frac{6400}{2 \times 9.81} = \underline{326.2 \text{ m}}$$

You should appreciate that this answer must be considered as highly theoretical, because we have not included any effect of air resistance, which would be quite considerable in such a case. The actual maximum height reached would be substantially less than the figure calculated. But you have to learn to walk before you learn to run, and the mathematics required to allow for the effect of resistance of the air would be far too complex for us at this stage.

Frame 33 contains a further eight 'drill' examples which you should be able to attempt. Do not forget the reminder at the end of the previous frame—do not divide your problem into two parts. This applies particularly in the cases of Examples 3, 5 and 6. Examples 7 and 8 are more complex; it is not just a matter of selecting a formula and substituting. You will need a strategy for solving problems of this type, and you may need some help with these two.

33

'Drill' exercises: motion under gravity

1. A stone falls from the top of a cliff of height 150 m. What will be its velocity after 2 seconds? How far will it have fallen in 4 seconds? How long will it take to reach the cliff foot? What then will be its velocity?
   Ans. 19.62 m/s; 78.48 m; 5.53 s; 54.25 m/s.
2. A body is projected vertically upwards with an initial speed of 40 m/s. Calculate the time elapsed for the speed to fall to 10 m/s, the time for it to reach its maximum height, and the maximum height.
   Ans. 3.058 s; 4.077 s; 81.55 m.
3. A missile is projected vertically upwards from the top of a tower of height 30 m with an initial velocity of 27 m/s. How long will it take to reach the ground?
   Hint: $x =$ *minus* 30 m.
   Ans. 6.453 s.
4. Careful measurement shows that the time of fall of a stone from the top of a tower is 2.4 seconds. Calculate the height of the tower, and the final velocity of the stone.
   Ans. 28.25 m; 23.544 m/s.
5. A missile is fired vertically upwards from a gun with an initial velocity of 160 m/s. Calculate the maximum height it will reach, the time to reach this height, and the time to reach a height of 500 m.
   Ans. 1304.8 m; 16.31 s; 3.5 s and 29.12 s.
6. A projectile fired vertically upwards is observed to reach a maximum height of 120 m. Calculate its initial velocity and the time elapsed before it returns to the ground.
   Ans. 48.52 m/s; 9.89 s.
7. A projectile is fired vertically upwards with an initial velocity of 80 m/s. A second projectile is fired exactly one second later from the same point with an initial velocity of 100 m/s. Calculate the velocity of each projectile when they are both at the same height. Calculate this height, and the time elapsed from the instant of release of the first projectile, to the instant they attain the same height.
   Ans. 45.47 m/s up; 75.28 m/s up; 220.8 m; 3.52 s.
8. A projectile is fired vertically downwards from the top of a tower with an initial velocity of 20 m/s. At the same instant, a second projectile is fired vertically upwards from the base of the tower with an initial velocity of 50 m/s. Calculate the time elapsed before they

meet, the height at which they meet, and the respective velocities. Calculate the height and velocity of the second projectile when the first one reaches the ground. The height of the tower is 300 m.
Ans. 4.286 s; 124.2 m; 62.04 m/s down; 7.95 m/s up; 123.05 m; 9.25 m/s down.

34

We now come to the kinematics of bodies which rotate instead of moving along straight paths: bodies such as rotating wheels and electric motors. This type of motion will call *angular motion* (as distinct from the *linear* motion of bodies moving along straight paths). When an electric motor is switched on, it attains a certain speed in a certain time. Clearly there is some displacement—the rotor moves. There is also velocity—the rotor is spinning. And finally, there must be acceleration—the rotor's speed has changed. But we cannot make use of the same methods of analysis used for linear motion. To begin with, we cannot measure the displacement of the rotor in metres, because although the motor armature is obviously moving, it does not move bodily—that is, the armature remains in one place, and turns about a fixed point. Our problem reduces to measuring displacement in some other fashion.

This is not difficult. We already have two familiar methods of measuring angular displacement: the degree and the revolution. In order to come to a decision on a suitable method of measuring angular displacement, we shall examine the relationship between linear and angular motion. Although our motor armature does not move bodily, a point on its circumference moves; it moves in a circular path about the centre of rotation. Furthermore, such a point will move with a certain linear speed (so many metres per second) at any instant. And if we choose an instant during the time the rotor is speeding up, such a point will have also a specific value of linear acceleration. This same argument can be applied also to any other point on the armature; it does not apply only to points on the circumference. In fact, the only points to which it does not apply are points on the axis of rotation. The distance travelled, the speed of travel and the acceleration are all dependent on the radius of the point chosen. The bigger the radius, the more the distance travelled. (It is salutary to remember that as a consequence of the rotation of the earth, each of us travels approximately 25000 kilometres each day in a circular path about the earth's axis, at a speed of approximately 300 m/s. Even this pales into insignificance when it is realised that our motion along the earth's orbit round the sun is accomplished at the spanking pace of about 30 kilometres per second.)

So we can begin by considering the motion of a point P along a circular track.

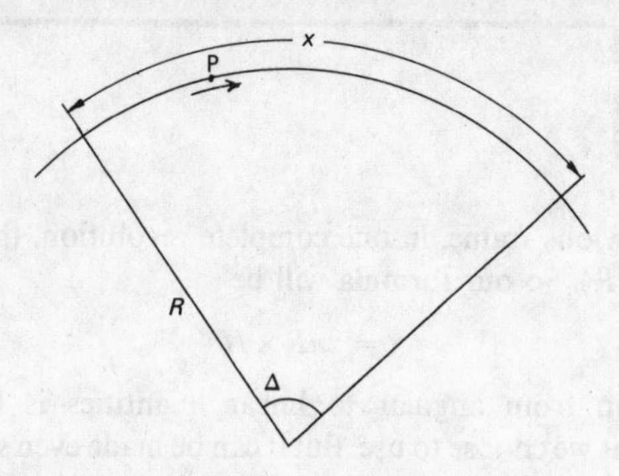

The point P moves a distance x around the circle, whose radius is R. The arc, of length x, subtends an angle of Δ degrees. Calculate x in terms of Δ and R.

35

If the wheel had made one complete revolution, of 360 degrees, the point P would have travelled the full circumference of the circle, and the corresponding distance travelled would have been $(2\pi R)$. So, for an angle of Δ degrees, the distance x will be given by

$$x = 2\pi R \times \frac{\Delta}{360}$$

You can use this simple formula to calculate that the distance around the earth corresponding to one degree of angle, assuming the earth's radius to be 3958 miles, is 69.08 miles. The distance corresponding to one minute of angle (one-sixtieth of a degree) is therefore 1.1513 miles, or 6080 feet. This distance is called the nautical mile, and a ship's speed is always stated in nautical miles per hour, called knots.

Now suppose that instead of defining the angle in degrees, we define it in revolutions. Derive a formula relating x to R and the number of revolutions, N (N, of course, can be less than 1).

36

As in the previous frame, in one complete revolution, the value of x would be $(2\pi R)$. So our formula will be

$$x = 2\pi R \times N$$

So conversion from angular to linear quantities is not difficult, whichever unit we choose to use. But it can be made even simpler. Both our formulae contain the terms (2π). If we change our unit of angular displacement to an angle consisting of $(1/2\pi)$ revolutions, or approximately 57.3 degrees, then for one complete revolution, as before, the distance covered will be $(2\pi R)$. So for one unit of this new angular unit, which we call the *radian*, the corresponding distance is

$$x = \frac{(2\pi R)}{2\pi} = R$$

This indicates another way of defining a radian of angular displacement. The radian is the angle subtended by an arc equal to 1 radius. The general formula relating linear displacement to angular displacement now becomes

$$x = R \times \theta$$

where θ is now the angle, *in radian measure.*

Thus, the adoption of what at first appears to be a rather odd and unfamiliar unit of angular measure results in a very simple formula relating linear and angular displacement. And we shall see in the following frame that the same simple relationship applies also to linear and angular velocity, and to linear and angular acceleration.

In Frame 2, we introduced the concept of Dimensions, and the necessity for an equation of mechanics (or for that matter, an equation of any other field of science) to be correct as to dimensions. The equation

$$x = R \times \theta$$

may appear at first sight to contradict this principle. The dimension of x, the displacement, is clearly L. Similarly, R, a radius, also must have the dimension L. Then how can the equation balance when L on the right-hand side is multiplied by something else? Well, it does, because the dimensions of θ, an angle, do not exist; it is said to be dimensionless. Whether an angle is measured in degrees, or radians, or even whole revolutions, it is defined as a numerical fraction of a complete rotation—that is a circle. If you are changing from the old imperial system of feet and inches to the metric system, you have to throw away your foot rule, and buy a metre rule, but you can still use your protractor; a degree is one-three-hundred-and-sixtieth of a circle in any system of units, and a radian is a full circular angle divided by 2π.

Since the distance around the circumference of a circle is $2\pi \times R$, it follows that an angle of one radian includes a portion of circumference of this distance divided by 2π—that is R. This gives us another definition of a radian; the angle subtended at the centre of a circle by a circumferential distance of one radius. And in general

$$\text{angle in radians} = \frac{\text{arc of circumference}}{\text{radius}}$$

that is

$$\theta = \frac{x}{R}$$

which is a transposition of the formula above.

This is the time to remind you that henceforward, in all calculations involving angular measure, you must *always* express angular quantities in *radians*.

37

So we know now that the angle θ (in radians) subtended by a circular arc of radius R and length x is given by

$$\theta = \frac{x}{R}$$

$$\therefore x = \theta \times R$$

Linear and angular velocity can be similarly related. Imagine the point P in the diagram of Frame 34 to be moving with a linear velocity v. The radius from the centre to P will then sweep across at an angular velocity which we shall call ω (omega) radians per second.

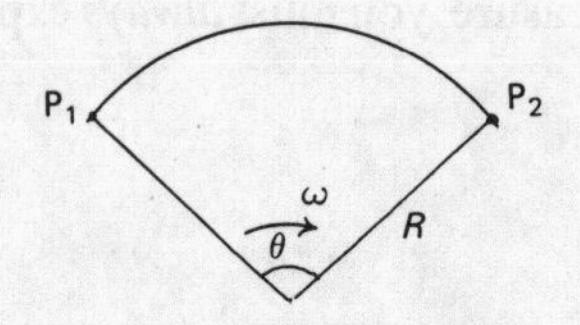

$$\text{Angular velocity } \omega = \frac{\text{displacement}}{\text{time}} = \frac{\theta}{t}$$

$$\text{Linear velocity } v = \frac{x}{t}$$

$$\text{But } x = \theta \times R$$

$$\therefore v = \frac{\theta \times R}{t} = \omega \times R$$

Now suppose that point P moves from P_1 to P_2 at an increasing speed, being v_1 at P_1 and v_2 at P_2. It will have a linear acceleration a given by

$$a = \frac{v_2 - v_1}{t}$$

The angular speed of the radius joining P to the centre will correspondingly increase in magnitude: that is, there will be an angular acceleration. We shall call this acceleration α (alpha). The units will be (radians per second) per second, or, rad s^{-2}. We can calculate the angular speeds ω_1 and ω_2 at points P_1 and P_2

$$\omega_1 = \frac{v_1}{R}; \quad \omega_2 = \frac{v_2}{R}$$

Substituting for v_2 and v_1 in the expression above for a

$$a = \frac{\omega_2 R - \omega_1 R}{t}$$

$$= R\left(\frac{\omega_2 - \omega_1}{t}\right)$$

$$\therefore \; a = \alpha \times R$$

You must be quite clear that the acceleration a is the linear acceleration due to the change of speed along the circular path. The direction of this acceleration (recalling that acceleration is a vector quantity) is either forwards or backwards along the track, according to whether the speed is increasing or decreasing; either way, its direction is tangential to the track. We shall see later that there will also be a centripetal acceleration, which arises out of the change of *direction* of the velocity of P, as it moves round the circular track.

38

The three formulae are repeated here, by way of summary.

$$x = \theta \times R$$

$$v = \omega \times R$$

$$a = \alpha \times R$$

This is a good point to stop and remind you that angular measurement in radians is a convenient unit for purposes of calculation, and is an essential if the above formulae are to be used. Ordinary people dealing with rotating machinery do not need to use the unit; indeed, it is probable that most people have never heard of it. Car drivers understand each other perfectly when speaking of engine speeds of 6000 revs per minute. A machine operator is no doubt quite capable of drilling twelve holes in a plate, on a circle at intervals of 30 degrees, and he might well down tools and take steps to initiate a strike if his drawing called for twelve holes spaced at an interval of 0.5236 radians. The circular scale on his machine would in any case be engraved in degrees and subdivisions thereof. So from now onwards, whenever you have to deal with angular motion in your work, remember that the units of angular displacement, velocity and acceleration must always be expressed in radians.

Since angular speed is very frequently expressed in revolutions per minute, it will be convenient to have a formula to hand for converting this into radians per second, or for converting the answer to a calculation in radians per second back to revolutions per minute. Calling the speed in revolutions per minute N, and the speed in radians per second ω, see if you can obtain such a formula.

39

$$\omega(\text{rad/s}) = \frac{N\ (\text{rev/min})}{60} \times 2\pi$$

So the required formula is

$$\omega = \frac{2\pi N}{60}$$

Conversion from degrees to radians and the reverse is often necessary. Since each revolution of 360 degrees contains 2π radians, the conversion formula will be

$$\theta_{rad} = \theta_{deg} \times \frac{2\pi}{360} = \theta_{deg} \times \frac{\pi}{180}$$

Have a go at the following simple conversions; the working is completed in Frame 40.

(a) Convert $1\frac{1}{2}$ revolutions to radians.
(b) Convert 60 degrees to radians.
(c) Convert 4 radians to revolutions.
(d) Convert 3.1416 radians to revolutions.
(e) Convert 120 rev/min to rad/s.
(f) Convert 8.56 rad/s to rev/min.
(g) Convert 0.00007272 rad/s to rev/day.

40

Conversions

(a) $\theta_{rad} = \theta_{deg} \times \frac{\pi}{180} = (1\frac{1}{2} \times 360) \times \frac{\pi}{180} = \underline{9.425 \text{ rad}}$

(b) $\theta_{rad} = \theta_{deg} \times \frac{\pi}{180} = 60 \times \frac{\pi}{180} = \underline{1.047 \text{ rad}}$

(c) $\theta_{deg} = \theta_{rad} \times \frac{180}{\pi} = 4 \times \frac{180}{\pi} \theta_{rev} = \theta_{deg} \div 360 = \underline{0.637 \text{ rev}}$

(d) $\theta_{deg} = \theta_{rad} \times \frac{180}{\pi} = 3.1416 \times \frac{180}{\pi} \theta_{rev} = \theta_{deg} \div 360 = \underline{0.5 \text{ rev}}$

(e) $\omega = \frac{2\pi N}{60} = \frac{2\pi \times 120}{60} = \underline{12.566 \text{ rad/s}}$

(f) $N = \omega \times \frac{60}{2\pi} = 8.56 \times \frac{30}{\pi} = \underline{81.742 \text{ rev/min}}$

(g) $N = \omega \times \frac{30}{\pi} = 0.00007272 \times \frac{30}{\pi} \text{ rev/min}$

$$= 0.00007272 \times \frac{30}{\pi} \times 60 \times 24 \text{ rev/day}$$

$$= \underline{1.00 \text{ rev/day}}$$

(The last answer is clearly the speed of revolution of the earth.)

41

Let's return to the three formulae listed at the beginning of Frame 38. Here they are again

$$x = \theta \times R$$

$$v = \omega \times R$$

$$a = \alpha \times R$$

For an application of these formulae, consider the following problem.

Problem A simple hoist consists of a load raised by a rope which is wound on to a drum of diameter 2.5 m. The load starts from rest and is raised with constant acceleration a distance of 8 m in 4.5 seconds. Determine (a) the number of revolutions the drum makes; (b) the drum speed at the end of the 4.5 seconds; (c) the angular acceleration of the drum.

Attempt this problem yourself first; do not work from first principles when solving part (a) but use the first of the three formulae. And before attempting (b) and (c) you may have to go back to Frame 13 and revise your linear kinematics. That is, calculate the velocity and acceleration of the load, and then use the second and third of the three above formulae. The complete solution is found in the following frame.

42

The distance moved by the rope will be the same distance moved by a point on the rim of the drum.

(a)

$$x = \theta \times R$$

$$8 = \theta \times (\tfrac{1}{2} \times 2.5)$$

$$\therefore \theta = 6.4 \text{ rad} = \frac{6.4}{2\pi} \text{ rev} = \underline{1.019 \text{ rev}}$$

(b) We require the velocity of the load after $4\frac{1}{2}$ s (see Frame 13)

$$x = \tfrac{1}{2}(u+v)t$$

$$8 = \tfrac{1}{2}(0+v)4\tfrac{1}{2}$$

$$\therefore v = \frac{2 \times 8}{4\frac{1}{2}} = 3.556 \text{ m/s}$$

The velocity of the rope will be the same as the velocity of the rim of the drum.

$$\omega = \frac{v}{R} = \frac{3.556}{(\frac{1}{2} \times 2.5)} = \underline{2.845 \text{ rad/s}}$$

(c) We require the acceleration of the load.

$$x = ut + \tfrac{1}{2}at^2$$

$$\therefore 8 = 0 + \tfrac{1}{2}a(4\tfrac{1}{2})^2$$

$$\therefore a = \frac{8 \times 2}{(4\frac{1}{2})^2} = 0.790 \text{ m s}^{-2}$$

$$\alpha = \frac{a}{R} = \frac{0.790}{(\frac{1}{2} \times 2.5)} = \underline{0.632 \text{ rad s}}$$

43

Here is a second problem.

Problem A car has a uniform acceleration. It passes one check-point at 10 m/s and a second check-point 12 seconds later at 34 m/s. Calculate the number of revolutions made by the wheels between the two points, the angular acceleration of the wheels, and the angular speeds of the wheels at the two points. The wheel radius is 0.34 m and it is assumed that the wheels do not slip.

The problem is very similar to the previous one, although it may not appear so at first sight. But if you imagine a car standing on a moving belt, with the wheels turning at such a speed that the car itself is stationary, it becomes clear that the distance moved by the belt is the same as the peripheral distance moved by the wheel rim. The belt velocity and acceleration will also be the same as the velocity and acceleration of the wheel rim. Now to bring the belt to rest, the car would have to move at the speed of the belt in the opposite direction, with an acceleration equal to that of the belt, again in the opposite direction. So our simple linear/angular conversion formulae apply.

The solution is *not* given for this problem, but the answers you should obtain are: 123.58 revs; 5.88 rad s^{-2}; 29.41 rad/s; 100 rad/s.

44

We now need a set of formulae for the solution of problems of angular motion similar to the four we obtained for linear motion, starting at Frame 9. We can use the same arguments to obtain these formulae, and we shall find that they will be in all respects exactly analogous to the earlier ones. We begin, then, by imagining a body rotating with angular acceleration (assumed constant) and we draw a graph of angular velocity against time. We shall call the initial velocity ω_0 and the acceleration α. Our graph will look like this;

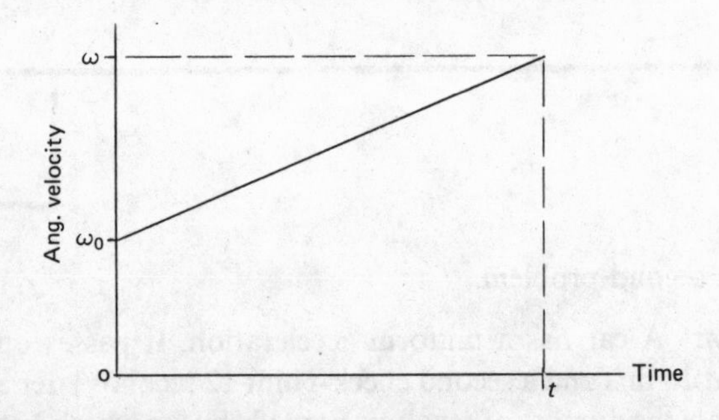

Derive an expression for the angular velocity ω after a time t. The reasoning is exactly the same as in Frame 9 in case you have trouble.

45

α is the increase of angular velocity in unit time. So the increase in time t will be (αt). So the final velocity will be $\omega_0 + (\alpha t)$. So the formula will be

$$\omega = \omega_0 + \alpha t \qquad (1)$$

We need not repeat the argument of Frame 8 to show that the area under a graph of angular velocity against time will give the displacement. So make use of this principle to obtain the second formula.

46

Displacement = area under velocity–time graph

$$\theta = t \times \tfrac{1}{2}(\omega_0 + \omega)$$

or
$$\theta = \tfrac{1}{2}t(\omega_0 + \omega) \qquad (2)$$

We can alternatively divide the area into a triangle and a rectangle.

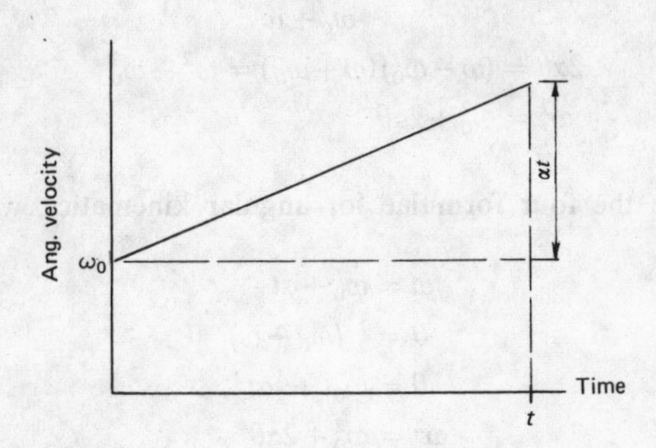

The increase of angular velocity in time t will be (αt). Express the area as the sum of a rectangle and a triangle to obtain Formula (3).

47

The rectangular area is $(\omega_0 \times t)$ and the triangle is $(\tfrac{1}{2} \times t \times (\alpha t))$. Adding

$$\theta = \omega_0 t + \tfrac{1}{2}\alpha t^2 \qquad (3)$$

Take any two of the formulae and eliminate t to obtain the last formula.

48

From formulae (1) and (2)

$$\omega = \omega_0 + \alpha t$$

$$\theta = \tfrac{1}{2}t(\omega_0 + \omega)$$

$$\therefore\ t = \frac{\omega - \omega_0}{\alpha} = \frac{2\theta}{\omega_0 + \omega}$$

$$\therefore\ 2\alpha\theta = (\omega - \omega_0)(\omega + \omega_0) = \omega^2 - \omega_0^2$$

$$\therefore\ \omega^2 = \omega_0^2 + 2\alpha\theta \qquad (4)$$

Here are the four formulae for angular kinematics, written out together

$$\omega = \omega_0 + \alpha t \qquad (1)$$

$$\theta = \tfrac{1}{2}t(\omega_0 + \omega) \qquad (2)$$

$$\theta = \omega_0 t + \tfrac{1}{2}\alpha t^2 \qquad (3)$$

$$\omega^2 = \omega_0^2 + 2\alpha\theta \qquad (4)$$

You can see that all four formulae are exactly analogous to the four formulae for linear kinematics, and have been derived in the same manner. For the solution of problems of linear motion, it is necessary, or at least very desirable, to remember all four formulae. But provided you can remember these, you can use them to write down, at sight, the corresponding four formulae for angular motion, without the extra trouble of having to remember four more formulae.

This second group of formulae can, of course, be derived using the calculus, in the same way that we did in Frame 14, but we have not bothered to do this.

One word of warning. Although the linear and angular formulae are analogous, they are not, term for term, identical as to dimensions. Linear displacement has the dimension of length, but angular displacement is dimensionless. (This is clear from the formula at the beginning of Frame 37: the dimensions of θ are $|x/R|$ which is Length $\div$ Length.) You can show for yourself that the dimensions of angular velocity will be T^{-1} and angular acceleration T^{-2}.

Problems of angular kinematics tend to be generally simpler than those of linear kinematics, and a single example (given in the next frame) should be sufficient exercise for the present.

49

Problem A large machine operates on the following 3-stage cycle.

(a) Starting from rest and increasing to a working speed of 15 rev/min in 45 seconds.
(b) Running at working speed for 4 minutes.
(c) Shutting off and coming to rest with an angular retardation of 0.008 rad s^{-2}.

The machine is required to be overhauled after each complete 5000 revolutions. How many cycles of operation may be allowed between overhauls? If the cycles are repeated continuously without a break between, what will be the time between overhauls?

The solution follows in Frame 50. But meanwhile, try to solve this by yourself. Remember the following points.

1. Sketch the velocity–time graph.
2. For each part of your solution, list what you have been told, and also what you want, and choose the formula which connects these quantities.
3. Wherever possible, avoid making use of an answer calculated in one part for the solution of another part.
4. Remember: *radians*, not revs!

50

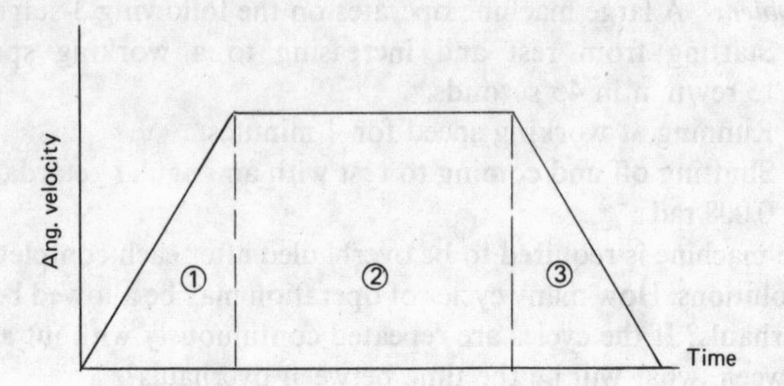

We can call the three stages 1, 2 and 3, the three corresponding times t_1, t_2 and t_3, and the corresponding angular displacements θ_1, θ_2 and θ_3. For the first part of the problem, it is clear that we require θ_1, θ_2 and θ_3.

Stage 1: use Formula (2) (Frame 48)

$$\theta = \tfrac{1}{2}t(\omega_0 + \omega)$$

$$\theta_1 = \tfrac{1}{2} \times 45\left(0 + \frac{15 \times 2\pi}{60}\right) \text{ rad}$$

$$= \tfrac{1}{2} \times 45 \times \frac{\pi}{2} \times \frac{1}{2\pi} \text{ rev} = 5.625 \text{ rev}$$

Stage 2: simply

$$\theta_{\text{rev}} = 4 \times 15 = 60 \text{ rev}$$

Stage 3: Formula (4)

$$\omega^2 = \omega_0^2 + 2\alpha\theta: \quad 0 = \left(\frac{15 \times 2\pi}{60}\right)^2 + 2(-0.008)\theta_3$$

$$\therefore \theta_3 = \left(\frac{\pi}{2}\right)^2 \times \frac{1}{2 \times 0.008} \times \frac{1}{2\pi} \text{ rev} = 24.54 \text{ rev}$$

$$\therefore \theta_1 + \theta_2 + \theta_3 = 90.17 \text{ rev}$$

$$\therefore \text{ number of cycles} = \frac{5000}{90.17} = 55.45 \text{ cycles, that is, } \underline{55 \text{ cycles}}$$

$t_1 = 45$ s; $t_2 = 4 \times 60 = 240$ s

Stage 3: Formula (1)

$$\omega = \omega_0 + \alpha t$$

$$0 = \frac{15 \times 2\pi}{60} - 0.008 t_3$$

$$\therefore\ t_3 = \frac{\pi}{2 \times 0\,008} = 196.3\ \text{s}$$

$$t_1 + t_2 + t_3 = 481.3\ \text{s}$$

$\therefore$ time between overhauls $= 55 \times 481.3\ \text{s} = \underline{2656.5\ \text{s}}$

51

We stated in Frame 2 that velocity was a vector quantity. In our work up to this point, this has not been of significance, but now we have to take into account this special property of velocity. When a body is moving around a circular path, it is undergoing an acceleration. This does not mean that it must be increasing or decreasing speed (although it may be doing that as well). Even though the speed around the path may be constant, the body is still accelerating, and this type of acceleration is called *centripetal acceleration.*

It must be realised that, being a vector, velocity changes not only if its magnitude changes, but also if its direction changes. Now when a body travels round a curved path, its direction is continually changing, even though its speed may remain constant, and therefore it is accelerating. If you feel inclined to object to this, you must realise that acceleration does not mean merely slowing down or speeding up; it means change of velocity, and a change of direction is just as significant as a change of magnitude. To evaluate a directional change of velocity, you must be able to sketch simple velocity vector diagrams. You may have had some practice in this sort of work previously, but it will not take long to go through it again. It is simpler to begin with displacement because this is easier to visualise. Suppose we add two displacements: 1 metre at 0 degrees and 2 metres at 90 degrees. What will be the resulting total displacement? Work it out and check in the next frame.

52

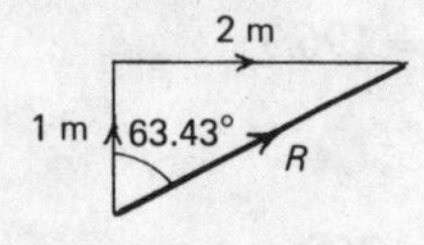

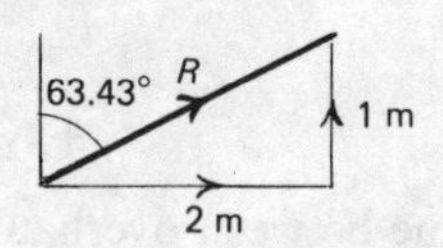

You may 'add' the displacements in either order; whichever way you do it, the resulting displacement R will be 2.236 metres at 63.43 degrees. Notice that the two displacement components are added 'end-to-end'—the arrows denoting direction must run the same way—or, more correctly, the tail of one vector must be attached to the head of the other. The following 'addition' is wrong

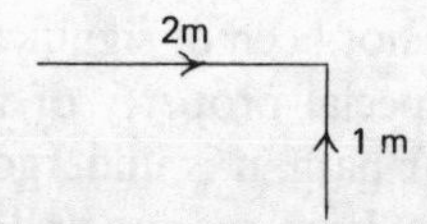

The arrow on the resultant R is in the opposite sense to the arrows on the components. This should be clear if you consider what we are trying to do. If you walk 1 m at 0° (due north) and then 2 m at 90° (due east) you will arrive at the same point that you would have if you had walked a distance of 2.236 m in a direction 63.43° east of north. If you reversed the arrow on the vector R, you would have a diagram representing the addition of three displacements which would bring you back to your starting-point.

Exactly the same rules apply to the addition of velocities. This is hardly surprising, as velocity is displacement per unit time. So, as a simple exercise, determine the resultant of two velocities, the first being 4 m/s at 0°, and the second 6 m/s at 270°. There is no need to draw accurately; a sketch will suffice. You should obtain an answer of 7.21 m/s at 303.7°. The next frame shows the working.

53

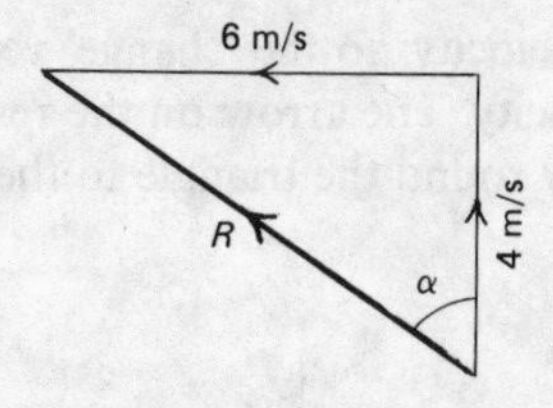

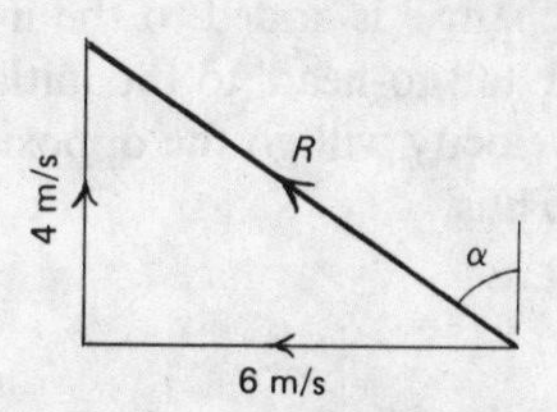

The length of the resulting vector R is 7.21 m/s. The angle α is 56.3°, so that the compass direction of R would be 303.7°. (This situation would arise if a ship sailed northwards in a sea with a westerly drift, or if an aeroplane flew in a cross-wind.)

Now let's turn the problem round. At a certain instant, a body has a velocity of 10 m/s at 90°. After 2 seconds, its velocity has changed to 10 m/s at 100°. What was the change of velocity?

Here, we are given the original velocity, and the final ('resultant') velocity, and we require to find the change. Use the simple statement:

$$\text{Initial velocity} + \text{change} = \text{final velocity}$$

and from the diagram representing this statement you should be able to show that the change of velocity was 1.743 m/s at 185°. The problem is solved in the following frame for you.

54

The 'change' is added to the initial velocity; so the 'change' vector is added 'tail-to-head' to the initial velocity. The arrow on the resulting new velocity will go the opposite way round the triangle to the other two. Thus

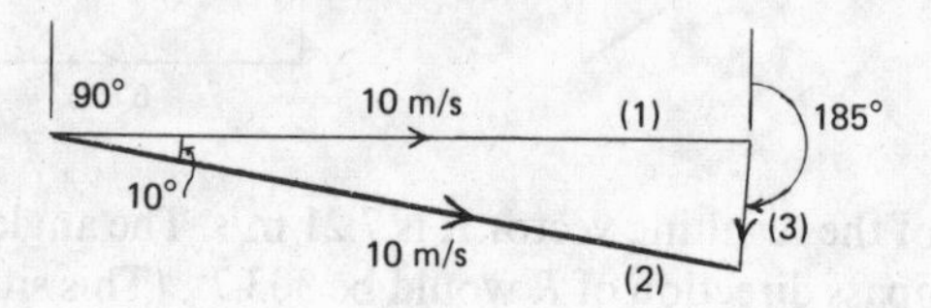

(1) is the initial velocity; (2) is the resultant, or final velocity. So (3) must be the change of velocity. Notice the importance of having the correct directions of the arrows. If you attempted this exercise yourself and got it wrong, the chances are that you mistakenly added the two velocities, 10 m/s at 90° and 10 m/s at 100°. Either by very careful scale-drawing and measurement, or by simple trigonometry, you should find that the change of velocity is 1.743 m/s at 185°. Since this change took 2 seconds, the average acceleration is

$$a_{av} = \frac{1.743}{2} = 0.872 \text{ m s}^{-2}$$

55

Now imagine a car travelling at 10 m/s around a circular track, and suppose that in 2 seconds it travels round a 10° arc of the track, thus

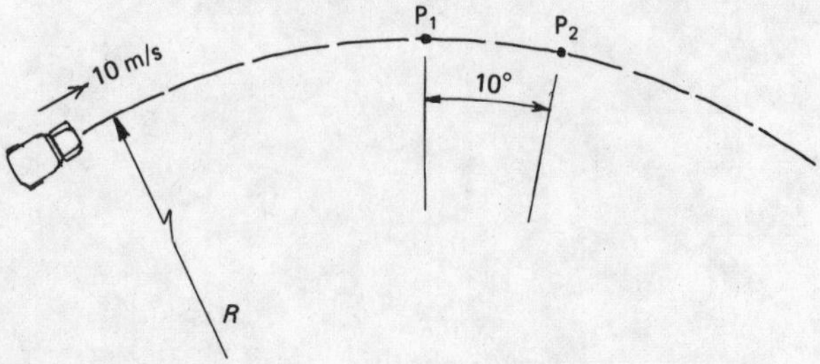

Draw the velocity vectors for the car at the two points P_1 and P_2 and thus determine the change of velocity which occurs during these 2 seconds. Remember: do not add the velocities; the velocity at point P_2 is the resultant of the velocity at point P_1 and the velocity change.

56

The vector diagram you should get is, of course, exactly the same as the one in Frame 54.

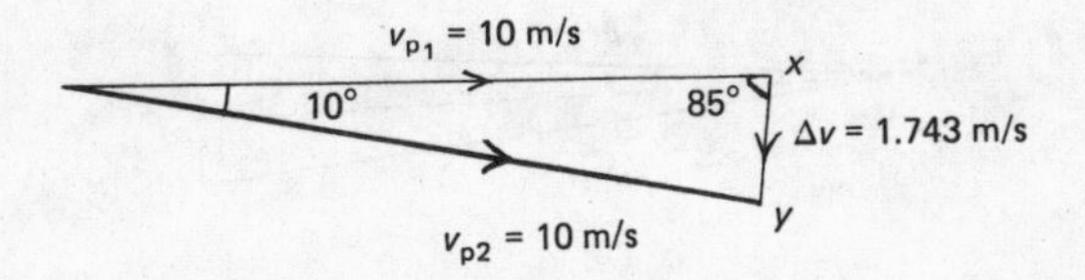

and the change of velocity, xy, will be 1.743 m/s at 85° to the direction of v_{P_1}. So the average acceleration of the car during this period will be 0.872 m s^{-2}.

But we do not want to have to draw diagrams of this sort every time we encounter motion along a circular track; neither do we want to know the average acceleration during a short period of time. What we require is the *instantaneous* acceleration, at a given point, at an instant of time. We can work round to this, firstly by assuming a track velocity v instead of 10 m/s, and secondly assuming an arc along the track of θ instead of 10°.

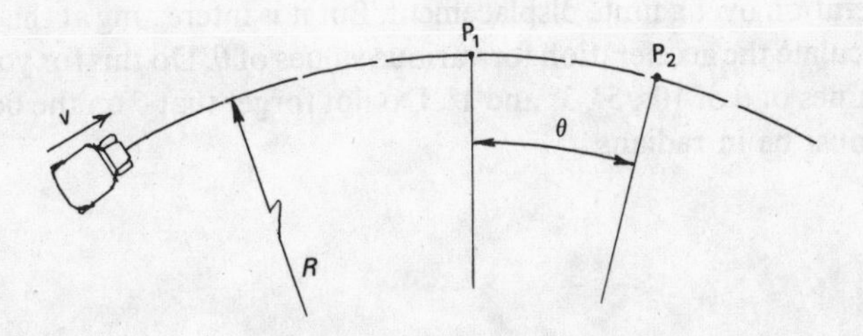

Sketch the velocity vectors for points P_1 and P_2 and try to calculate from your diagram (a) the change of velocity, (b) the time taken for this change, and (c) an expression for the average acceleration during this time.

57

Your diagram should look like this. It is, of course, very similar to the previous one in shape.

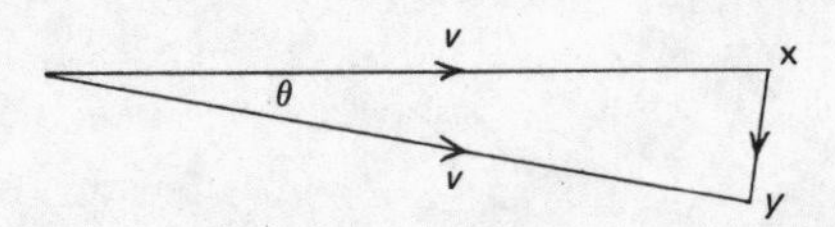

To determine xy we can construct a perpendicular

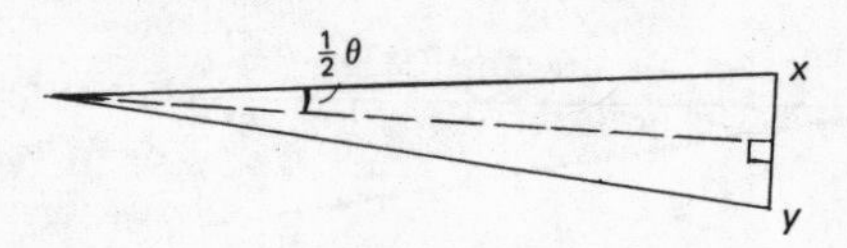

Then

$$xy = 2(v \sin(\tfrac{1}{2}\theta))$$

The distance P_1P_2 around the curve $= \theta \times R$ (not forgetting that θ is in radians). So the time taken will be $\theta R/v$. So average acceleration is

$$\frac{\text{change of velocity}}{\text{time}} = \frac{2v \sin(\frac{1}{2}\theta)}{\theta R/v}$$

$$= \frac{2v^2 \sin(\frac{1}{2}\theta)}{R\theta}$$

This is still not yet quite what is required; it is still the average acceleration over a finite displacement. But it is interesting at this point to calculate the acceleration for various values of θ. Do this for yourself for values of θ of 10°, 5°, 3° and 1°. Do not forget that θ on the bottom line must be in radians.

58

For $\theta = 10°$, the acceleration a is

$$a = \frac{2v^2 \sin 5°}{(10 \times \pi/180)} = 0.99873\frac{v^2}{R}$$

and you can show by similar calculations that the remaining accelerations are

$$\text{for } 5°,\ a = 0.99968\frac{v^2}{R};\ \text{for } 3°,\ a = 0.99989\frac{v^2}{R};\ \text{for } 1°,\ a = 0.99999\frac{v^2}{R}$$

It is quite clear from these figures that the closer to zero we make θ, the closer the coefficient of (v^2/R) approaches 1. We can see therefore that as θ approaches zero, the instantaneous acceleration, or the limiting value of the expression—in other words, the centripetal acceleration—is

$$a_0 = \frac{v^2}{R}$$

59

We can arrive at this expression for centripetal acceleration in another way, by considering a small sector of the circle

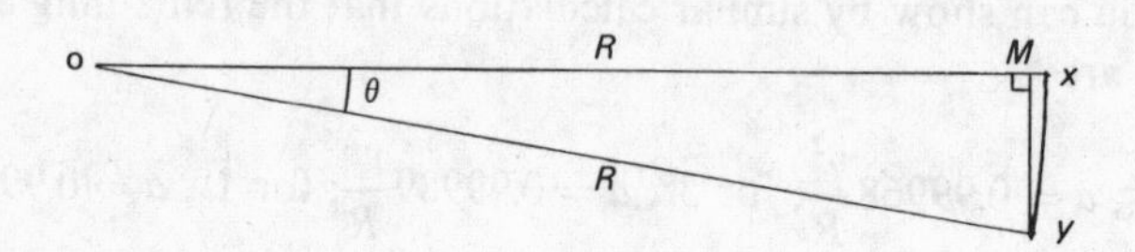

xy is the arc of the circle subtended by the small angle θ. Now if we construct a perpendicular from y to M, then we can say

$$\sin\theta = \frac{\text{yM}}{R}$$

Also, in radian measure

$$\theta = \frac{\text{xy}}{R}$$

But it is clear from the diagram, firstly that xy is very nearly the same length as yM (it is not easy to show the different lines even in this diagram), and secondly that the smaller we make the angle θ, the closer to each other the two lengths become. When angle θ is very small, it therefore becomes true to say

$$\theta \simeq \sin\theta$$

(remembering, as always, that θ is in radian measure). For example, sin 5° on a calculator is (to six places) 0.087156, and 5° converted to radians is, as you can check for yourself, 0.087266; a difference of approximately 0.1 per cent, and if you compare the sine and the angle in radians for 1° you will find a difference of approximately 0.006 per cent. This approximation for small angles is very common in mathematical analysis; you have almost certainly met it before. So if we now re-examine the expression in Frame 57 for average acceleration

$$a_{av} = \frac{2v^2 \sin(\frac{1}{2}\theta)}{R\theta}$$

we can say, because we assume θ to be very small

$$a_{av} \simeq \frac{2v^2 \times (\frac{1}{2}\theta)}{R\theta}$$

$$\simeq \frac{v^2}{R}$$

and although this is an approximation for small values of θ, it becomes exact 'in the limit' as θ approaches zero.

The diagram of Frame 57 shows the change of velocity, xy, to be almost perpendicular to the vector v, and the smaller we make θ, the closer it approaches perpendicularity. (In the limit, the isosceles triangle has an apex angle of zero and equal base angles of 90°.) The conclusion is that the direction of the centripetal acceleration is perpendicular to the direction of the tangential velocity, and directed inwards, towards the centre of the path.

Using the relationship (see Frame 38) $v = \omega R$, we can express centripetal acceleration alternatively by

$$a_0 = \frac{(\omega R)^2}{R} = \omega^2 R$$

This form is convenient when dealing with rotating bodies such as wheels and rotors, whereas the first form is applicable to such problems as vehicles on circular tracks, planets revolving round suns, etc., where the linear speed of the body is more likely to be known, or required, than its angular speed.

60

We can summarise the work from Frame 51.

1. When a body travels with constant speed v on a circular path of radius R, it undergoes a centripetal acceleration of magnitude v^2/R.
2. When a body turns about an axis with a constant angular speed ω, a point within the body at a radius R undergoes a centripetal acceleration of magnitude $\omega^2 R$.
3. The centripetal acceleration is always directed towards the centre of curvature of the curved path.

61

Try this example before going on to the 'drill' examples at the end of the programme.

Problem A car travels at a constant speed of 80 km/hour around a circular track of mean radius 90 m. Calculate the magnitude of the centripetal acceleration. Given that the car wheels have a radius of 0.3 m, evaluate the centripetal acceleration of a point on the rim of a wheel, relative to the wheel centre.

The first part is a simple application of the first formula (Frame 58) but do not forget that all units must be converted to metres and seconds and their derivatives. For the second part, refer to Frame 38, second formula, and recall that the linear speed of a point on the rim of a wheel will be the same as the speed of the car (assuming there is no slip of the wheel on the road).

62

$$a_0 = \frac{v^2}{R}; \quad v = \frac{80 \times 1000}{60 \times 60} = 22.22 \text{ m/s}$$

$$\therefore \ a_0 = \frac{22.22^2}{90} = \underline{5.486 \text{ m s}^{-2}}$$

$$\omega = \frac{v}{R_{\text{wheel}}} = \frac{22.22}{0.3} = 74.07 \text{ rad/s}$$

$$\therefore \ a_0 = \omega^2 R = (74.07)^2 \times 0.3 = \underline{1645.9 \text{ m s}^{-2}}$$

This example serves to show that wheels and rotors which turn at high speeds can undergo very high values of acceleration. This becomes important when we examine the forces required to produce these accelerations.

This concludes the work of this programme. Frame 63 comprises 'drill' exercises on angular motion and centripetal acceleration, and Frame 64 concludes with some general revision examples for the whole programme.

63

'Drill' exercises: angular motion; motion in a circular path

1. A wheel rotates at 150 rev/min. It comes to rest in 10 minutes with uniform retardation. Calculate this retardation, and the total number of revolutions made by the wheel in coming to rest.
   Ans. -0.0262 rad s^{-2}; 750 rev.
2. A vehicle has wheels of diameter 0.85 m. Calculate the angular velocity of the wheels when the vehicle travels at 24 km/hour. Also evaluate the centripetal acceleration of a point on the rim of a wheel. If the vehicle comes to rest in a distance of 28 m with uniform retardation, calculate the corresponding angular retardation of a wheel, and the number of revolutions made by a wheel in coming to rest.
   Ans. 15.69 rad/s; 104.6 m s^{-2}; 1.867 rad s^{-2}; 10.49 rev.
3. A motor rotates at 5000 rev/min. It slows to rest uniformly 10 seconds after being switched off. Calculate the angular retardation, and the number of revolutions made by the armature in coming to rest.
   Ans. 52.36 rad s^{-2}; 416.7 rev.
4. A design safety rule for high-speed cast-iron wheels states that the maximum linear speed of a point on the rim must be 1.6 km per minute. Use this rule to calculate the maximum speed, in rev/min, of wheels of diameter (a) 1.2 m, (b) 80 mm. In each case, calculate the centripetal acceleration of a point on the rim. Calculate the maximum allowable diameter of a wheel required to run at 3600 rev/min.
   Ans. 424.4 rev/min; 1185.2 m s^{-2}; 6366 rev/min; 17.778 m s^{-2}; 141 mm.
5. A flywheel is uniformly accelerated from rest to 1500 rev/min in 12 seconds. The diameter is 1.24 m. Calculate (a) the angular acceleration, (b) the angle turned through in reaching the final speed, (c) the maximum linear velocity of a point on the rim, (d) the tangential acceleration of a point on the rim, (e) the maximum centripetal acceleration of a point on the rim.
   Ans. (a) 13.09 rad s^{-2}. (b) 150 rev. (c) 97.39 m/s. (d) 8.12 m s^{-2}. (e) 15.298 m s^{-2}.
6. A wheel is accelerated uniformly from rest at 7 rad s^{-2} for 20 seconds. It then turns at constant speed for 150 seconds. It then comes uniformly to rest in a further 15 seconds. Calculate (a) the

maximum speed, in rev/min, (b) the total revolutions made, (c) the final retardation.
Ans. (a) 1337 rev/min. (b) 3732.2 rev. (c) 9.33 rad s^{-2}.

7. At a certain instant, a car is travelling round a circular track of mean radius 80 m with a speed of 25 m/s and a forward acceleration of 3.5 m s^{-2}. What is the total acceleration of the car at this instant? *Hint*: vector sum of forward and centripetal accelerations, which are at right-angles.
Ans. 8.5607 m s^{-2}.
8. The wheels of a car are 0.55 m diameter. The car starts from rest and attains a speed of 80 km/hour in 32 seconds. It travels at this speed for 45 seconds. It is then brought to rest uniformly at 2.4 m s^{-2}. Calculate the total time for the journey, and the number of revolutions of the wheels for each stage of the journey.
Ans. 86.26 s; 205.8, 578.8 and 59.5 rev.
9. Calculate the centripetal acceleration of (a) a point on the earth's surface at the equator, assuming the earth's radius to be 4000 miles, (b) the earth itself due to its motion around the sun, assuming a circular path of radius 93 million miles.
Ans. (a) 0.034 m s^{-2}; (b) 0.00594 m s^{-2}.

64

General revision example

1. A motor cycle and car start from the same point at the same time, and travel along the same straight track. The cyclist accelerates uniformly at 1.4 m s^{-2} to a maximum speed of 60 km/hour; the car accelerates uniformly at 0.6 m s^{-2} to a maximum speed of 80 km/hour. Calculate the time taken for the car to overtake the cyclist, and the distance travelled in this time.
Ans. 56.24 s; 838.3 m.
2. A vehicle travels from A to B a total distance of 2 km. It starts from rest at A with acceleration 0.6 m s^{-2} for a time t_1. It then travels at constant speed for a further time t_2. It finally comes to rest at B after an additional time t_3, having retarded at 1.8 m s^{-2}. The total time taken is 120 seconds. Calculate the three times, and the speed over the second part of the journey.
Ans. 34.32 s; 74.25 s; 11.43 s; 20.59 m/s.
3. Two points A and B on a straight track are 800 m apart. Two vehicles start at the same instant, one travelling from A to B and the

other from B to A. The one leaving A accelerates from rest at 0.5 m s^{-2} until it attains a maximum speed of 48 km/hour. The vehicle leaving B accelerates from rest at 2.4 m s^{-2}, reaching a maximum speed of 40 km/hour. Calculate the distance from A where the vehicles meet, and the time taken.
Ans. 369.5 m; 41.06 s.

4. A shell is projected vertically from a gun with an initial velocity of 250 m/s. It is required to explode at a height of 1700 m. Calculate the time at which the fuse should be set, so that the shell explodes on its way up, Neglect air resistance.
Ans. 8.08 s.
5. A projectile is fired vertically upwards from the top of a tower of height 40 m with an initial velocity of 80 m/s. At the same instant, a second projectile is fired vertically upwards from the base of the tower with an initial velocity of 100 m/s. Determine the time elapsed before they are at the same height, and calculate this height.
Ans. 2 s; 180.38 m.
6. A body is dropped from rest from the top of a tower of height 50 m. After 2 seconds, a projectile is fired vertically upwards from the base of the tower with an initial velocity of 20 m/s. Calculate the height at which they meet.
Ans. 12.45 m.
7. A satellite may be assumed to have a circular orbit around the earth at a radius (to the earth's centre) of 9000 miles. One complete orbit takes $4\frac{1}{2}$ hours. Calculate the speed of the satellite in m/s and the magnitude of the centripetal acceleration. If the satellite is subjected to a linear retardation of 0.014 m s^{-2}, calculate how many orbits it will make before its speed is reduced by 10 per cent, assuming its orbital path is unaltered.
Ans. 5.618 km/s; 2.179 m s^{-2}; 2.35 orbits.
8. The armature of an electric motor reaches a speed of 5000 rev/min in 10 s from rest. This speed is maintained for 30 s. The power is then switched off and the armature comes to rest in a further 90 s. Find the angular acceleration and retardation, and the number of turns that the armature makes during each phase of the motion. Calculate also the angular velocity during the second phase.
Ans. 52.4 rad s^{-2}; 5.82 rad s^{-2}; 416.7, 2500 and 3750 turns; 523.6 rad/s.

[illegible] from [illegible] the [illegible] [illegible] [illegible] [illegible] [illegible]
[illegible] [illegible] [illegible] [illegible] of [illegible] [illegible]
[illegible] [illegible] [illegible] [illegible] [illegible] [illegible]
[illegible] [illegible] [illegible] [illegible] [illegible] [illegible]
[illegible] [illegible] of [illegible] [illegible]

[illegible] [illegible] [illegible] [illegible] [illegible] [illegible] of
[illegible] [illegible] [illegible] [illegible] [illegible] [illegible] [illegible]
[illegible] [illegible] [illegible] [illegible] [illegible] [illegible]
[illegible] [illegible] [illegible]

[illegible] [illegible] [illegible] [illegible] [illegible] the [illegible] [illegible] of
[illegible] [illegible] [illegible] [illegible] [illegible]
[illegible] [illegible] [illegible] [illegible] [illegible] the [illegible] of the
[illegible] [illegible] [illegible] [illegible] [illegible] [illegible]
before the [illegible] of the [illegible] [illegible] its height [illegible]
[illegible] [illegible]

[illegible] [illegible] from [illegible] the [illegible] [illegible] of the [illegible]
[illegible] [illegible] [illegible] [illegible] [illegible] [illegible] the [illegible]
[illegible] [illegible] [illegible] [illegible] [illegible] the height
[illegible] [illegible]

[illegible]

A [illegible] [illegible] [illegible] [illegible] [illegible] [illegible] [illegible]
[illegible] [illegible] [illegible] [illegible] [illegible] [illegible]
[illegible] [illegible] [illegible] [illegible] of the [illegible] [illegible] the
[illegible] [illegible] of the [illegible] [illegible] of the [illegible] [illegible]
[illegible] [illegible] [illegible] [illegible] [illegible] [illegible] [illegible]
[illegible] [illegible] speed [illegible] [illegible] of [illegible] [illegible]
[illegible] [illegible]

[illegible]

[illegible] [illegible] [illegible] [illegible] [illegible] [illegible] [illegible]
[illegible] [illegible] [illegible] [illegible] [illegible] [illegible]
[illegible] [illegible] [illegible] [illegible] [illegible] [illegible]
[illegible] [illegible] [illegible] [illegible] of the [illegible] of [illegible]
[illegible] [illegible] [illegible] [illegible] of the [illegible] [illegible]
[illegible] [illegible] [illegible] [illegible]
[illegible] [illegible] [illegible] [illegible]
[illegible]

Programme 3: Elementary Kinetics

1

We may as well begin with a definition. *Kinetics* is the study of the effects of forces on the motion of bodies; the whole of this programme is devoted to this topic. Now it is probably clear that such a topic covers a considerable area of work. (Even if you hadn't realised this, the thickness of this particular programme would be a sufficient indication. And this is only 'Elementary Kinetics'.) Nevertheless, it will be helpful to appreciate at the beginning that the principles of kinetics are simple, and that all of the work stems from Newton's Second Law of Motion. There are various ways of stating this Law, but the following will suffice for us.

> "The acceleration of a body is proportional to the resultant force acting on it, and the direction of the acceleration will be the same as the direction of the resultant force"

We shall come back to this in the next frame, but for the purpose of this introduction it will be sufficient to state it, and to draw some conclusions from it. Consider first the word 'resultant'. You will know already that it is possible for a body to have several forces acting on it, and yet not be in a state of acceleration. When this is so, the resultant of the forces acting on the body is zero; in other words, the body is in a state of equilibrium. A kinetic problem therefore usually begins with an examination of the forces acting on a body, to determine whether a resultant exists, and if so, what its value is, and what its direction is. Putting it another way, the study of *kinetics* requires a previous thorough study of *statics*.

Now consider the word 'acceleration'. This is defined as a change of the velocity of a body. A body subjected to a force system which is not in equilibrium will have its velocity changed—either in magnitude, or in direction, or in both. The corollory is that if the velocity of a body is known to be changing, then it must be subjected to a resultant force. In order that velocity changes can be perceived and calculated, a previous knowledge of kinematics is necessary. Thus, the study of *kinetics* requires a previous study of *kinematics*.

So, before you become too deeply immersed in this present programme, you are reminded that you should have already worked through the two programmes 'Elementary Statics' and 'Elementary Kinematics'. In Statics, you will particularly require to be fluent in the technique of force resolution.

Having got this advice over, you may now put this text away until you have dealt with the earlier ones, or, if you are in the happy position of having already done this, you may turn over to Frame 2.

2

Understanding Dynamics is not and never can be, a mere learning of formulae. Nevertheless, a formula can do much to shorten and simplify the work of analysis. It is our job in this frame to reduce Newton's Second Law to a simple and workable formula.

Experimental work at the time of Newton showed that the acceleration of bodies subjected to forces varied according to two things. Firstly, acceleration was seen to vary directly with the force; that is, double the force on a body produced double the acceleration, and so on. (You may have seen this yourself, using the apparatus known as Fletcher's Trolley.) Secondly, it was seen that some bodies 'resisted acceleration' more than others; the same force acting on two bodies produced different values of acceleration. Bodies possessed a property called Inertia; this was the property of 'staying put', so to speak, when acted on by a force. Inertia was not merely a matter of size. A football, for example, will resist the action of a kick far less than a sphere of concrete of the same size—as I hope you are prepared to believe without proof! Neither is it directly a property of weight. Both the football and the concrete sphere might be weightless out in space, but if you tried to kick them both, you would damage your foot on the concrete just the same as if you were earthbound. We call this property of resisting force Mass (which is really just another word for inertia) and classical experiment (such as Fletcher's Trolley) shows that acceleration is inversely proportional to mass (that is, the same force acting on double the mass will produce half the acceleration, and so on).

Combining these two relationships, we can now write a formula

$$a = K \times \frac{F}{m}$$

where a is the acceleration of a body, F is the resultant force acting on it, m is its mass, and K is a constant.

Now we come to units of measurement. With acceleration, using the S.I. system, we are committed to the use of the metre for length and the second for time. One unit of acceleration is thus one metre per second per second, or, briefly, m s^{-2}. Mass is measured in kilograms, and this is an arbitrary unit. By this, we mean that its value does not depend on anything else that we have previously defined, as acceleration depends on length and time (both of which are arbitrary units). One kilogram of mass is defined as the mass of a piece of platinum kept at the

International Bureau of Standards near Paris. (Originally, a kilogram was defined as the mass of 1000 cubic centimetres of water, but this definition is now obsolete.)

At this point, we *define* the unit of force, which we so choose, to make the constant K in our formula equal to 1: we define unit force as the force required on one kilogram mass to cause an acceleration of 1 m s^{-2}. This unit of force, as you probably know, is called the *newton* (abbreviated to N). Committing ourselves to these units, we rewrite our formula

$$a = \frac{F}{m}$$

and we may bring this frame to a close by re-arranging the formula in its most general and familiar form, without fractions

$$\Sigma F = m \times a$$

the purpose of the Σ (Greek capital letter 'sigma') being to remind ourselves that the force we are concerned with is the *resultant* of a system of forces.

3

Confusion sometimes arises because people have become accustomed to measuring force in kilograms. You know, of course, that the weight of a body is the gravitational pull exerted on it by the earth. A body of mass 10 kg is 'heavy' because the earth pulls it downwards—with a force of 10 kg, we would say. Newton's Law of Gravitation tells us that, provided a body is close to the earth, its weight is directly proportional to its mass. Thus, it has come to be established that weighing a body is a convenient and accurate way of measuring its mass, and, ordinarily, there is no real need to distinguish between weight and mass. I am not seriously concerned whether the mass of my body is 75 kg, or whether I weigh 75 kg. When I buy 2 kg of potatoes, it is of no importance to me whether the mass of the potatoes is 2 kg or whether the earth attracts them with a force of 2 kg. But we have begun the study of kinetics with a formula, and a definition of the unit of force; our formula requires that force shall be measured in newtons, not kilograms. Now it is a fact of experimental observation that, again provided they are close to the earth's surface, all bodies fall with the same acceleration (if the effects of air resistance are eliminated); this value of acceleration is about 9.81 m s^{-2}.

Now if a body having a mass of exactly 1 kg falls downwards with an acceleration of 9.81 m s^{-2} due only to the downward pull of its own weight, that weight must be 9.81 newtons (since a force of 1 newton, by definition, would produce an acceleration of 1 m s^{-2}). Thus, the weight W, in newtons, on earth, of any body of mass m kilograms, can be simply calculated from the formula

$$W = m \times g$$

wherein g has the particular value of 9.81 m s^{-2}. (Remember that this formula relating weight and mass applies only in cases where the body is fairly close to the earth's surface. Do not make the mistake of using it to calculate, say, the weight of a satellite orbiting the earth at a height of 35,000 km. Even at the modest height of about 200 km, there would be a measurable reduction of weight.)

The whole of this programme will be based on this system of units: mass will always be in kilograms (kg), length will always be in metres (m), time will always be in seconds (s), and force will always be stated, and will be calculated, in newtons (N). This forms part of the S.I. system; the system has gained very wide acceptance, and most textbooks now use it. Nevertheless, you should always be aware of other systems of units, and should not be nonplussed if you encounter problems in which the data are not given in S.I. units. The cgs system, for example, takes the gram as the unit of mass, and the centimetre as the unit of length. (The second remains as the unit of time.) In this system, unit force—the force to accelerate 1 gram at 1 cm s^{-2}—is called the dyne (which, as you can show for yourself, has a value of 10^{-5} newtons). Also, in some older English textbooks, you may come across the pound as unit mass and the foot unit of length, the second remaining the unit of time. Since g in this system is 32.2 ft s^{-2}, unit force, called the poundal, is (1/32.2) times the weight of the pound mass.

To summarise so far, problems of kinetics are approached using the general formula

$$\Sigma F = m \times a$$

In this formula

ΣF is the resultant force on the body, measured in newtons (N)
m is the mass of the body, measured in kilograms (kg)
a is the acceleration of the body, measured in metres per second per second (m s^{-2}).

We can now begin to solve problems using our general formula

$$\Sigma F = m \times a$$

Here are three bodies, each subjected to simple force systems.

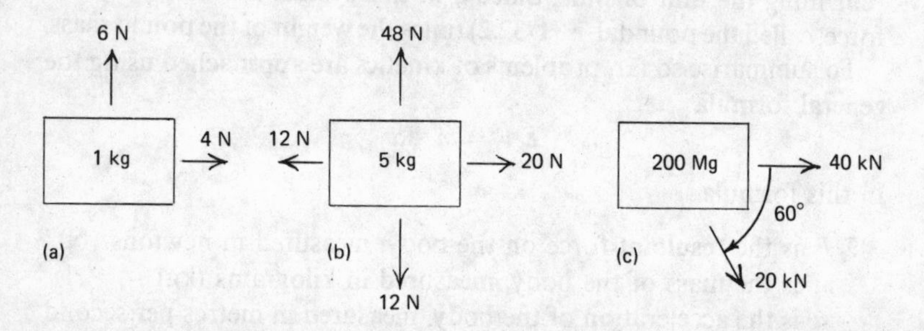

Calculate the acceleration for each of these three bodies. Find the value only; do not bother about its direction. You can assume that the forces shown comprise all the forces acting on the body. So you should begin by calculating the magnitude of the resultant force, and then substitute in the formula. The answers are

$$\text{(a) } 7.21 \text{ m s}^{-2}; \text{ (b) } 7.38 \text{ m s}^{-2}; \text{ (c) } 0.2646 \text{ m s}^{-2}$$

and if you have difficulty arriving at these answers, a bit of revision of your statics might be necessary; refer to 'Elementary Statics', Frames 13 to 17. In any case, the solutions are given in the next frame.

5

(a)

6 N, 4 N, R

$$\text{Resultant } R = \sqrt{4^2 + 6^2} = 7.21 \text{ N}$$

$$\Sigma F = m \times a$$

$$\therefore 7.21 = 1 \times a$$

$$\therefore \underline{a = 7.21 \text{ m s}^{-2}}$$

(b)

48 N, 12 N, 20 N, 12 N $\equiv$ 36 N, 8 N, R

$$R = \sqrt{36^2 + 8^2} = 36.88 \text{ N}$$

$$\Sigma F = m \times a$$

$$\therefore 36.88 = 5 \times a$$

$$\therefore \underline{a = 7.38 \text{ m s}^{-2}}$$

(c)

40 kN, 60°, 20 kN $\equiv$ 40 kN, 20 cos 60° kN, 20 sin 60° kN $\equiv$ 50 kN, 17.32 kN, R

$$R = \sqrt{50^2 + 17.32^2} = 52.92 \text{ kN}$$

$$\Sigma F = m \times a$$

$$\therefore 52.92 \times 10^3 = (200 \times 10^3)a$$

$$\therefore \underline{a = 0.2646 \text{ m s}^{-2}}$$

6

I hope you were able to remember the rules for combining forces which we covered in 'Elementary Statics'. You now begin to see, I hope, the importance of being able to determine the resultant of a number of forces. The three examples of the last frame were very simple; certain forces were shown acting on a body, and all you had to do was to calculate the magnitude of the acceleration. You might have noticed that no stipulations were made as to the size of the body, or the locations of the forces acting on it; the forces were treated as though they all passed through one point. When this is done, the body is said to be treated as a *particle*. Later, we shall consider the location of the forces when we look at the dynamics of *rigid bodies*.

If we think now about some real engineering problems, as distinct from the three simple examples of Frame 4, we find two important differences. Firstly, in real problems, nobody tells you what all the forces are; you have to find out for yourself what forces are acting on a body. We shall look at this aspect shortly. Secondly (and this time, this is an advantage), the nature of the problem is usually such that we should know something about the direction of the acceleration. To give a simple example: if we are examining the dynamics of a train on a track, or a vehicle on a road, we know that the direction of the acceleration must be along the track, or along the road; the train or the vehicle is not likely to start digging into the earth, or cruising up into the air! To be a little more accurate in this statement: we may not know the *direction* of the acceleration, but we do know its *line of action*. Later examples will make this quite clear. In Frame 7, we shall consider carefully all the forces which we are likely to encounter when analysing the kinetics of a body.

7

So far as this elementary text is concerned, all our bodies will be on, or near to, the earth, and therefore, they will be acted upon by their *weight*, which will act vertically downwards. Next, if a body lies on a surface (for example, a road or a rail), the surface will exert a force on the body. When the surface is smooth, this force will have a direction perpendicular to the surface—we call it a *reaction.* When the surface is not smooth, the reaction will be inclined to the perpendicular, and in such cases, we resolve the force into a perpendicular component and a tangential component, which latter we call a *friction* force. And we recall that, generally, a friction force acts in such a direction as to prevent, or partially prevent motion. Now if a body has a rope or a wire tied to it (for example, the cage of a lift), the rope or the wire exerts a pull, or a *tension* force, on the body. A tension—not a thrust; you cannot tie a rope on to a body and use the rope to push it away from you. If, on the other hand, a stiff rod or bar is attached to a body (such as a towing-bar on a trailer), this rod can exert either a *tension* or a *thrust* on the body. (When you are climbing a hill, towing your camping-trailer behind you, the tow-bar will most likely be in tension, but if you are going downhill with the brakes on, it will almost certainly be in compression, exerting a thrust on the car. It is, indeed, for this very reason that you have a bar, and not a rope. With a tow-rope, there would always be the possibility that the trailer could move up and collide with the back of the car.) For a powered vehicle, there will always be a driving force from the engine—called the *tractive force.* We must analyse this in detail in the next frame. Finally, if the body is moving in a fluid—water or air—this may exert a force on the body, by means of *fluid pressure.* There are many other forces, such as magnetic and electrostatic forces, but they are not very significant in mechanical problems.

8

By way of preparation for problem-solving, stop at this point, and think very carefully exactly *how* a car engine drives the car along the road. What is absolutely necessary for the engine to be effective (apart from the obvious answer, 'petrol')? And while on this subject, how does a propeller-driven aircraft work? And a jet-plane? And a steamship? All these are examples of power-driven vehicles, which are propelled by a tractive force. This is provided indirectly by the engine, but in order to analyse such a vehicle kinetically, we must know the actual mechanism whereby the engine provides the tractive force. You may not know the answers to these questions, but it will help you if you try and work out for yourself what is happening before reading on.

9

You will probably know that the car engine connects to the driving-wheels (by means of a gearbox and a clutch). You certainly know that the driving wheels rest on the ground. The wheel axles are given a forward twist, or torque, by the engine. But because the wheels are on the ground, when the axle is twisted, the bottom of a driving wheel pushes backwards against the road. At this stage, you recall Newton's Third Law: "Every force produces an equal and opposite reaction." If the wheel is pushing backwards on the road, *the road is pushing forward on the wheel.* It is this forward push of the road on the wheel which constitutes the real tractive force. It now becomes clear, I hope, that what is absolutely necessary for the engine to be effective is *friction.* You may have experienced the exasperation resulting from trying to drive a car along an icy road; the engine is in perfect order, but cannot be used to propel the car. Students often find this confusing, because they have been taught earlier that "friction opposes motion." But in the case of a wheel-driven vehicle, it is friction that is necessary to produce the motion.

A propeller-driven aircraft operates differently; the propeller pushes backwards against the air. So, again, we invoke Newton's Third Law; the air must push forward against the aircraft. A consequence of this is that such craft operate less efficiently at high altitudes, where the air density is less. A jet-plane, in contrast, pushes a mass of effluent backwards at very high velocity; the engine thrust is the equal and opposite reaction force exerted by the effluent on the aircraft. This is independent of the density of the air; thus, a jet-plane operates more effectively at high altitudes, because the resistance of the air to the aircraft's motion is less as the density reduces. A steamship operates in the same manner as a propeller-driven aircraft, except that the propeller pushes against water instead of air.

We shall end this frame with a check-list of all the forces we are likely to meet when examining the motion of bodies: the forces discussed in Frame 7.

1. Weight
2. Surface normal reaction
3. Surface friction reaction
4. Rope or wire tension (pull only)

5. Bar tension or compression (pull or push)
6. Tractive force
7. Fluid pressure (including wind or water resistance).

10

Let's look at a problem now—a slightly more realistic one than the simple exercises of Frame 4.

Example

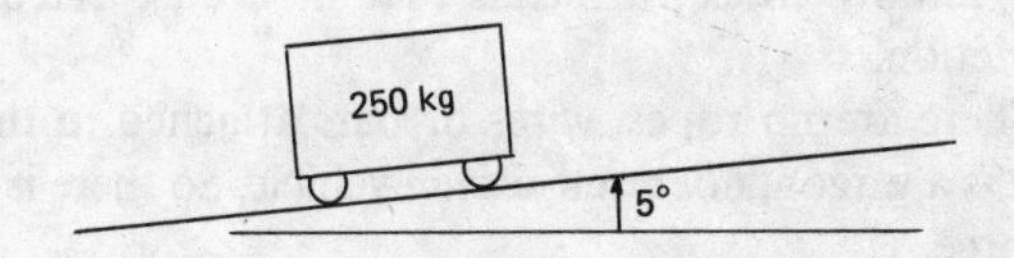

A wagon of mass 250 kg rests on a smooth rail having a slope of 5° as shown. Find its acceleration.

Look back to the check-list at the end of the previous frame. We have to find *all* the forces acting on the wagon. How many do you reckon? Put down all the forces you think we need. Then read on to Frame 11.

11

1. √ Weight. Obviously the wagon has weight; its value is ($250g$) N.
2. √ Surface normal reaction. The wagon rests on the rail; therefore the rail must exert a force on the wagon (we do not know its value).
3. × Surface friction. No mention is made of friction in the statement of the problem. If we were required to take it into account, we should have been given information about its value. In any case, when a problem refers to a 'smooth rail', or a 'smooth track', it means that we are expected to neglect friction.

4, 5. × There are no ropes, wires or bars attached to the wagon.

6. × It is a wagon, not a self-driven vehicle. So there is no tractive force.
7. × As the wagon moves, it will encounter resistance due to the air. But since we are given no information about the magnitude of the resistance, we shall be expected to assume it to be negligible—that is, zero.

The next stage is to draw a simple diagram of the wagon, showing the two forces that we have decided are acting on it. This diagram is called a *free-body diagram.* Its purpose is to show all the forces which are seen to be acting on the body, correct as to magnitude, if known, and direction, and also to show the line of action, and the direction (if known) of the acceleration. From here forward, *always* begin the solution of a kinetic problem by drawing the free-body diagram. In this particular problem, such a procedure may seem trivial and unnecessary, but as problems increase in complexity, you will begin to see how essential this practice is.

The free-body diagram looks like this

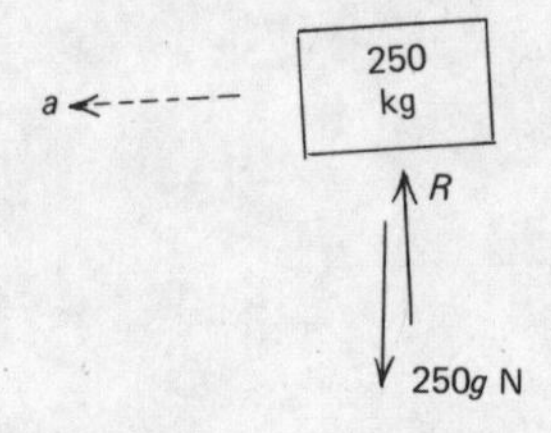

Notice: only the body and the forces are shown, not the track.

12

For a very simple diagram, there are several things that we have to say about this. Firstly, if a force is known to exist but we do not know its value, we give it a letter. In this case, we call the normal reaction force R. Secondly, it must be obvious that the line of action of the acceleration must be along the direction of the track. (As we pointed out earlier, it cannot be up into the air or down into the track.) And in this particular case, the acceleration must be down the slope, not up it. We shall meet problems later in which we shall not be able to state with certainty which way the acceleration acts. In such cases, we shall have to guess. Finally, notice the arrow representing the acceleration; this is dotted, deliberately to distinguish it from the two force-arrows. It is simpler to use a pen of a distinguishing colour for forces—I always use a red pen for forces—but this is not possible in a printed text.

Now, we have two forces acting on the body, and we know that the acceleration is directed down the 5° slope. The resultant of the two forces must therefore also be directed down the slope. We could construct a force 'polygon' such that the resultant is in the correct direction, thus

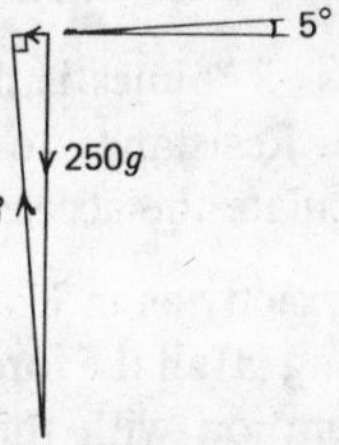

5°
250g
R

But it is much simpler to make use of the principle of force resolution. We can resolve the forces along two directions, *one being the direction of the acceleration and the other being perpendicular to it.* Thus

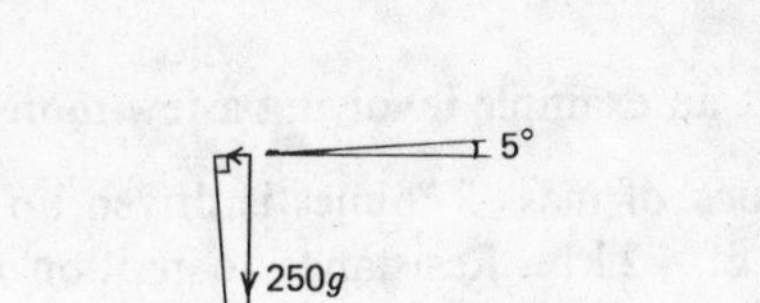

250g sin 5°
a
250 kg
R
250g cos 5°

We now write the general equation of motion $\Sigma F = m \times a$, having regard to the direction of a which is known to be down the slope. The equation is

$$250g \sin 5° = 250 \times a$$

from which

$$a = g \sin 5° = \underline{0.855 \text{ m s}^{-2}}$$

and you will note that we have arrived at the answer without having had to determine the force R. If we were required to find this, we could write an *equation of equilibrium* in the direction perpendicular to the acceleration

$$R - 250g \cos 5° = 0$$

giving a value for R of 2443.2 N. An equation of equilibrium may be considered as a special case of the equation $\Sigma F = m \times a$ when the value of acceleration a is zero: thus, $\Sigma F = 0$.

13

Now let's look at an example involving a few more forces.

> *Example* A truck of mass 2 tonnes is driven up a 12^0 slope by a tractive force of 4.2 kN. Resistance to motion due to the air is estimated at 450 N. Calculate the acceleration of the truck.

Your approach should be exactly as in Frame 11. Sketch a free-body diagram of the truck, showing (a) all the forces and (b) the acceleration. Do not confuse the acceleration with the force. (Why not use my method of reserving a red pen to indicate forces?) The line of action of the acceleration is clearly along the direction of the slope, and you may assume it to be up the slope. (We will come back to this point at the end of the solution.) Go through the force check-list carefully (Frame 9). You should find four forces. By the way, you should know that 1 tonne is 1000 kg.

14

Your free-body diagram should look like this

R
4200 N
a
2000 kg
450 N
12°
2000g N

Compare this carefully with your attempt and find out where you went wrong (if at all). When you have sorted things out, go ahead and write the equation of motion. Do not look at Frame 15 first; have a go yourself, and then check in Frame 15.

15

The equation is

$$4200 - 450 - 2000g \sin 12^\circ = 2000\,a$$

and since a is the only unknown term, we can determine it. The calculation is

$$a = \frac{(4200 - 450 - 2000 \times 9.81 \sin 12^\circ)}{2000}$$

$$= \underline{-0.1646 \text{ m s}^{-2}}$$

How are we to interpret this negative answer? The problem states that the truck is driven *up* the slope, and the negative value for a suggests that the acceleration is *down* the slope. Well, of course, this is quite possible; the truck may be moving up the slope, but its speed is reducing. This example serves to illustrate the principle that if the direction of the acceleration is unknown, or at best, not immediately obvious, we may assume it, and if our assumption is incorrect, this will be shown by a negative value.

16

We need to clear up one point about resolution. In the two examples so far, we have resolved forces along the two directions, one being the direction of the acceleration, and the other being perpendicular to it. Do we have to do this? The answer is, "No, not necessarily." The choice of directions is a matter of convenience. Look back to Frame 11. Suppose that instead of resolving along the slope and perpendicular to it, we chose instead to resolve horizontally and vertically. After all, we have some justification for this; the force of $250g$ N is already vertical. We then have to resolve force R into vertical and horizontal components. ***But*** we must now remember that the acceleration a is also a vector, and this also must be resolved. So our diagram becomes

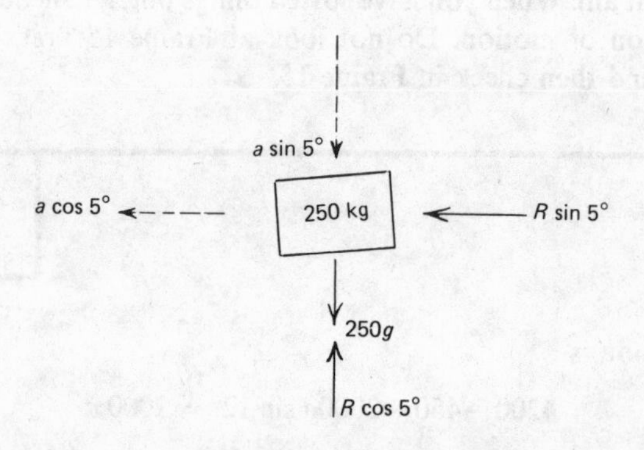

Now we have to write *two* equations of motion

$$R \sin 5^\circ = 250(a \cos 5^\circ) \quad (\Sigma F = m \times a \text{ horizontally})$$

$$250 - R \cos 5^\circ = 250(a \sin 5^\circ) \quad (\Sigma F = m \times a \text{ vertically})$$

and these two simultaneous equations must be solved, by eliminating R to find a. If you do this, you will end up with the same value for a which we found in Frame 12, but of course you will have had to do a lot more work. So for a general working rule, you should always choose the line of action of the acceleration as one of the directions for resolution.

Here is another simple example.

Example A locomotive with train has a total mass of 420 Mg. Calculate the required tractive force to drive it up a slope of 1 in 150, starting from rest and attaining a speed of 12 m/s in a distance of 50 m. Neglect any resistance to motion due to the air.

The problem is simple, but you need to brush up your elementary kinematics. See if you can solve it yourself. The answer is 632.3 kN. The solution is shown in the next frame. By the way, a slope of 1 in 150 always means (in this programme, anyway) that the *sine* of the angle of slope is 1/150.

17

You will most probably have seen that the acceleration is not stated, but that information is given to enable you to calculate it. The kinematic equation you need is

$$v^2 = u^2 + 2ax$$

Substituting

$$12^2 = 0 + 2 \times a \times 50$$

which gives a value for a of 1.44 m s^{-2}. This time, there is no doubt; the acceleration is definitely *up* the slope. Your free-body diagram is therefore

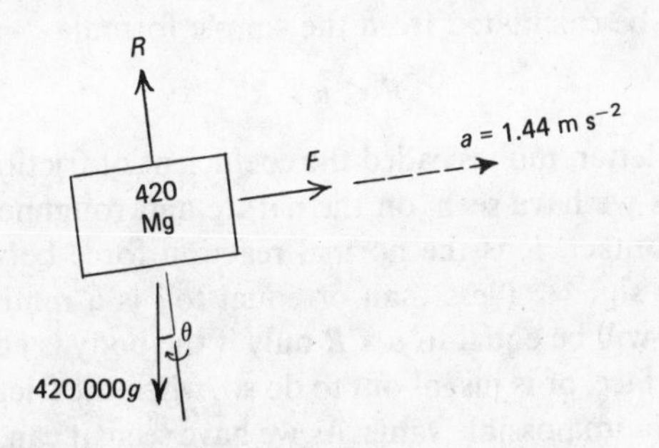

and the equation of motion is

$$F - 420{,}000g \sin\theta = 420{,}000a$$

$$\therefore\ F = 420{,}000 \times 1.44 + \frac{420\,000 \times 9.81}{150}$$

$$= \underline{632.3 \text{ kN}}$$

18

Let us introduce friction forces now. You need to recall what are sometimes called the 'Laws of dry friction'. For our purposes, these can be condensed to two.

1. Friction always acts in a direction that tends to oppose motion.
2. The friction force is proportional to the normal reaction force between the surfaces in contact.

Friction is a reaction force: it comes into existence only when another force tries to move a body. If a wooden block is at rest on a horizontal table, with no other forces acting upon it but its own weight, and the equal and opposite upward reaction force of the table, a friction force will not exist. But if in addition, a small sideways force is applied to the block, insufficient to move it, the table will exert an equal and opposite force, so that equilibrium is maintained. The table can do this because it is not perfectly smooth, but has tiny bumps and hollows on its surface which interfere with similar bumps and hollows on the undersurface of the block. If the sideways force on the block is steadily increased, the block will eventually begin to move, because the sideways friction force between table and block has a limited maximum value (determined by the natures of the two surfaces in contact, and the normal, or perpendicular, reaction force between them). This limited maximum value, F, can be calculated from the simple formula

$$F \leqslant \mu \times R$$

μ (the Greek letter 'mu') is called the coefficient of friction. Its value is dependent, as we have seen, on the nature and roughness of the two surfaces in contact. R is the normal reaction force between the two surfaces. The sign $\leqslant$ ('less than or equal to') is a reminder that the friction force will be equal to $\mu \times R$ only if the body is actually sliding across the surface, or is just about to do so, when the friction force will have its maximum possible value. As we have seen, it can, of course, be less than this value.

Friction forces are an area of mechanics where accuracy is difficult to achieve. A coefficient of friction cannot be calculated: it can only be determined experimentally. Tables of friction coefficients are to be found in textbooks and engineering handbooks, and may be used in order to calculate forces in particular circumstances. But a wise calculator will assume a wide possible margin of error—say, ± 20 per cent at least—in his calculations. Another phenomenon complicates the picture. If we go back to our simple block on the table, and steadily increase the sideways force until the block begins to slide, we should in all probability find that it would also begin to *accelerate*, because it is found that in most cases, less force is needed to keep a body sliding than to start it sliding. This peculiarity has earned the engaging name of 'sticktion'. So engineers treat friction forces sceptically and with caution, and if it is necessary to know them with accuracy, they always resort to experiment.

Here is a simple friction example, which you should be able to solve unaided. Follow the rules: draw a free-body diagram and show all the forces. You should find three forces. The answer is 4.803 m s^{-2}.

> *Example* A body of mass 20 kg is placed on an inclined surface having a slope of 40°. The friction coefficient between body and surface of plane is 0.2. Calculate the acceleration of the body.

19

Here is the free-body diagram, and the working.

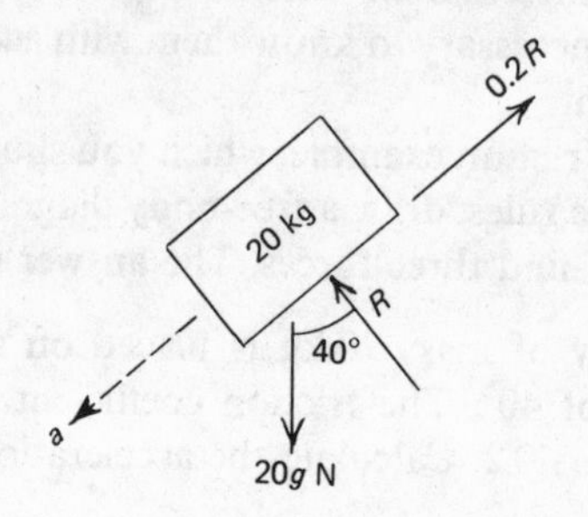

Down the slope: $20g \sin 40° - 0.2R = 20a$

Across the slope: $20g \cos 40° = R$

Substitute for R:

$$20g \sin 40° - 0.2(20g \cos 40°) = 20a$$

Cancel 20 and re-arrange

$$\therefore a = g(\sin 40° - 0.2 \cos 40°)$$
$$= \underline{4.803 \text{ m s}^{-2}}$$

I hope you had no difficulty in indicating the direction of the acceleration down the slope. Of course, it would not matter if you had shown it as acting up the slope; you would finish up with a negative value for a. But the friction force *must* be shown acting upwards; a body placed on a slope will tend to slide down, and friction will try to stop it. Be sure you understand and correct any mistakes you made.

20

Now we'll complicate the problem a little.

> *Example* A body of mass 30 kg is projected up an inclined plane of slope 35° with an initial speed of 12 m/s. The coefficient of friction is 0.15. Calculate (a) how far up the slope it travels before coming to rest, (b) how long it takes to reach the highest point, (c) how much longer it takes to return to its starting-point and (d) how fast it is travelling when it gets there.

The point to note here is that this is really two problems. When the body is moving up the slope, the friction force will act downwards. But when it begins to come down, the direction of friction reverses, and acts up the plane. So the acceleration of the body will be different coming down from what it was going up. Attempt the first part only to begin with. Draw a free-body diagram, and calculate the acceleration. You should find this to be 6.832 $m\,s^{-2}$ acting down the plane. Then you will need to recall your kinematic equations to find the distance and the time taken. The answers should be 10.538 m and 1.756 s.

21

Here is the free-body diagram and the working for the first part of the example.

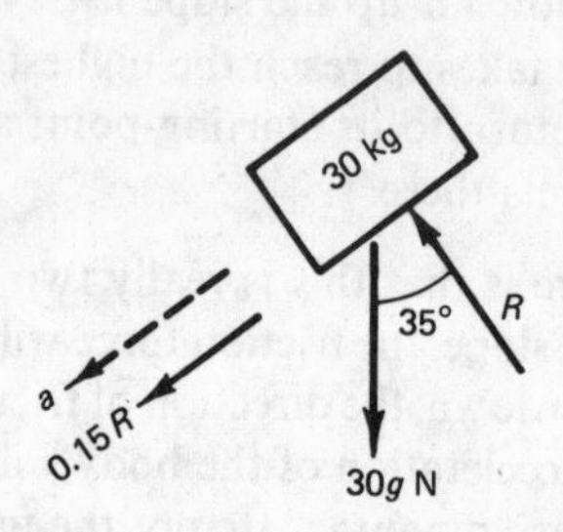

Down plane: $30g \sin 35° + 0.15\,R = 30\,a$

Across plane: $30\,g \cos 35° = R$

Substituting and re-arranging as for the previous example

$$30g \sin 35° + 0.15(30g \cos 35°) = 30a$$

$$\therefore\ a = g(\sin 35° + 0.15 \cos 35°)$$

$$= \underline{6.832\ \text{m s}^{-2}}$$

To find distance

$$v^2 = u^2 + 2ax$$

$$0 = 12^2 + 2(-6.832)\,x$$

$$\therefore\ x = \frac{144}{2 \times 6.832} = \underline{10.538\ \text{m}}$$

and the time

$$v = u + at$$

$$0 = 12 + (-6.832)\,t$$

$$\therefore\ t = \frac{12}{6.832} = \underline{1.756\ \text{s}}$$

Observe carefully (especially if you made mistakes) that when we apply the kinematic equations, we take the direction up the slope as positive; hence the acceleration must be reckoned negative.

The technique for solving the second part of the problem is the same, but of course the friction force now acts up the plane. The value of the acceleration should be 4.421 m s^{-2}, the time to descend to the starting-point 2.183 s, and the speed at the bottom 9.653 m/s. The distance travelled down the plane is, of course, the value of x calculated above.

Frame 22 shows all the working.

22

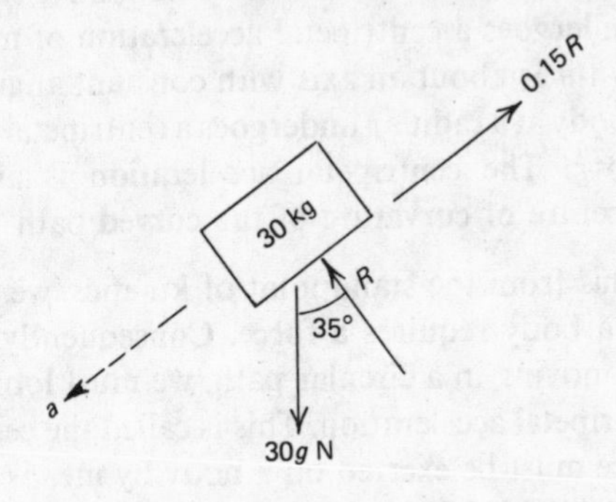

Down plane: $30g \sin 35° - 0.15R = 30a$

Across plane: $30g \cos 35° = R$

$$30g \sin 35° - 0.15(30g \cos 35°) = 30a$$

$$\therefore\ a = g(\sin 35° - 0.15 \cos 35°)$$

$$= \underline{4.421 \text{ m s}^{-2}}$$

The body begins to slide down the plane with this acceleration, and has to travel the distance $x = 10.538$ m.

To find time:

$$x = ut + \tfrac{1}{2}at^2$$

$$10.538 = 0 + \tfrac{1}{2}(4.421)t^2$$

$$\therefore\ t = \sqrt{\frac{2 \times 10.538}{4.421}} = \underline{2.183 \text{ s}}$$

To find final velocity:

$$v^2 = u^2 + 2ax$$

$$= 0 + 2 \times 4.421 \times 10.538$$

$$= 93.186$$

$$\therefore\ v = \underline{9.653 \text{ m/s}}$$

23

Before looking at the next type of problem, you might need to prepare yourself by referring back to 'Elementary Kinematics', Frame 51 and following. This deals with the motion of a particle along a circular path, and introduces the concept of centripetal acceleration. Recapitulating the essence of this work:

> "When a body travels with constant speed v on a circular track of radius r, it undergoes a centripetal acceleration of magnitude v^2/r. When a body turns about an axis with constant angular speed ω, a point on the body at a radius r undergoes a centripetal acceleration of magnitude $\omega^2 r$. The centripetal acceleration is always directed towards the centre of curvature of the curved path"

Looking at this from the standpoint of kinetics, we know that the acceleration of a body requires a force. Consequently, whenever we observe a body moving in a circular path, we must look for the force causing the centripetal acceleration. This is called the centripetal force. Centripetal force must be exerted on a body by means of some other body; by Newton's Third Law, the first body must therefore exert an equal and opposite reaction, and this reaction is called the centrifugal force. For example, if you stand in a vehicle which moves round a bend, you feel a push on your body, pushing you round the bend. But your body pushes in the opposite direction outwards against the vehicle. This reverse push of your body on the vehicle is the centrifugal force. Of course, when considering the forces acting *on your body*, the centrifugal force must not be included—only the centripetal force.

24

Imagine a small body of mass m at the end of a light string of length L, the other end of the string being attached to a fixed point; the body moves at a constant angular speed ω in a horizontal circle. Such an arrangement is called a conical pendulum.

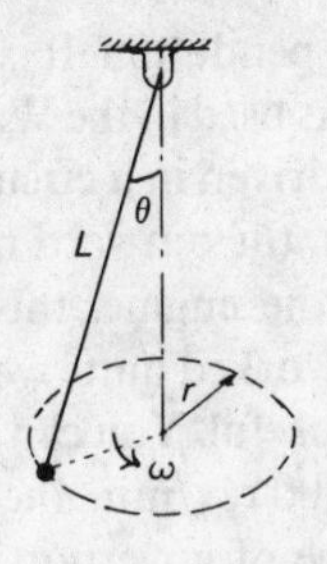

Since the body is moving in a circular path, it must have a centripetal acceleration directed towards the centre of rotation. This is shown on the free-body diagram of the body, when at the outermost part of the circle. The diagram also discloses the weight of the body, and the string tension, T, as the only two significant forces, air resistance being neglected.

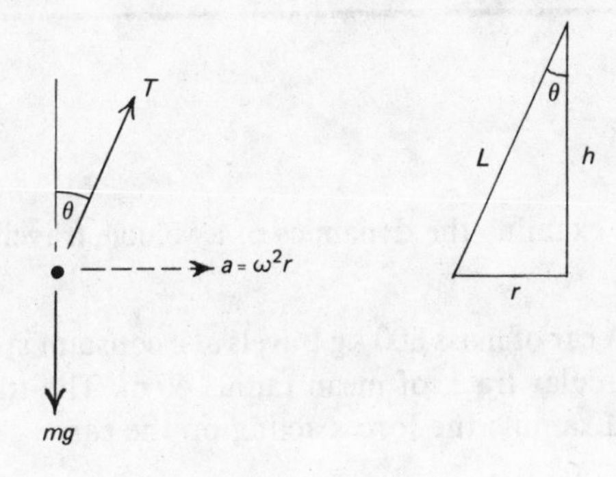

The diagram offers two equations

Horizontally, $\Sigma F = m \times a$:
$$\therefore\ T \sin\theta = m \times \omega^2 r$$

Vertically, $\Sigma F = 0$: $\therefore\ T\cos\theta = mg$

Eliminate T: $\therefore\ \tan\theta = \dfrac{\not{m}\,\omega^2 r}{\not{m}\,g}$

From the system triangle: $\tan\theta = \dfrac{r}{h}$

$$\therefore\ \frac{r}{h} = \frac{\omega^2 r}{g}$$

$$\therefore\ h = \frac{g}{\omega^2}$$

h is called the height of the pendulum. It is seen that h gets smaller as ω increases. This principle was used in the Watt engine governor; a ball on the end of a light rod was driven in a circular path by the engine shaft, and as the speed increased, the outward movement of the ball caused fuel to be shut off from the engine, thus limiting the speed. For a relatively slow engine this worked quite well, but as speeds began to get higher, the device was less useful. You can easily show for yourself that for a rotation speed of 300 rev/min the value of h would be only 9.9 mm. Such a simple type of governor is now a historical curiosity only, and has been replaced by more sophisticated devices, such as spring-loaded governors of the Hartnell type.

Notice that the actual centripetal force on the body is the horizontal component of the string tension. The other end of the string pulls outwards on the fixed string support; this is the centrifugal force.

25

Now we must examine the dynamics of a vehicle travelling around a circular track.

Example A car of mass 800 kg travels at a constant speed of 10 m/s around a circular track of mean radius 40 m. The track surface is horizontal. Examine the forces acting on the car.

Here is the complete free-body diagram, looking at the car end-on.

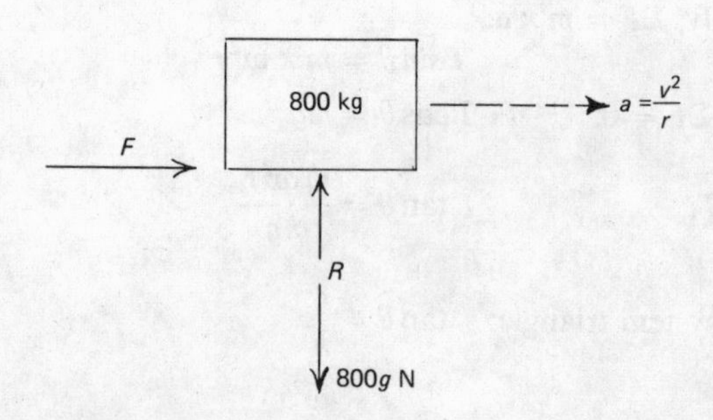

If you refer to your check-list in Frame 9, you can account for the downward weight of the car, and also for the vertical upward reaction R. We know that the car must have a centripetal acceleration. (The diagram is intended to represent the rear view of the car turning to the right.) So the acceleration must be horizontally to the right. A force F is shown, acting to the right. Now we know that there must be a right-directed force on the car, because it has a right-directed acceleration. But in drawing the free-body diagram, this is not sufficient evidence. Forces must be included because they can be explained in terms of the constraints on the body—that is the things which are actually touching it or exerting forces on it. So, although the acceleration tells you that there must be a force F to the right, you also have to justify this force, by stating exactly what is providing it. To put it another way, it is not sufficient to say, as learners are prone to do, "There must be a force because of the acceleration." This is to turn Newton's Second Law upside down. Forces *cause* acceleration; they are not caused *by* acceleration. So, having decided that there must be a force to the right to provide the acceleration which we know is there, the next question which you must answer is "What is actually providing that force?"

26

The answer to the previous question is "Friction". A car cannot travel round a curve on a flat road unless there is sufficient friction between tyres and road, as all car drivers will know—especially those who have tried to drive a car under icy conditions. We deduced in Frame 9 that a friction force is necessary to accelerate a car forwards or backwards. It is equally necessary here to accelerate it sideways.

Our free-body diagram discloses two equations

$$\text{Horizontally, } \Sigma F = m \times a: \quad \therefore F = 800 \times \frac{v^2}{r} = 800 \times \frac{10^2}{40} = 2000 \text{ N}$$

$$\text{Vertically } \Sigma F = 0: \quad \therefore R = 800\,g = 7848 \text{ N}$$

Let's find out in the next example how fast the car can go without slipping.

27

Example Calculate the greatest speed at which the car of the previous example (Frame 25) may travel around the circular track if the friction coefficient between road and tyres is 0.4.

We need not redraw the free-body diagram; the forces will be as previously, except that since the friction force must now have its maximum possible value, we replace F by μR. Friction force can always be less than μR but can never be greater. The speed v is now unknown. So the first equation of Frame 26 becomes

$$\mu R = 800 \times \frac{v^2}{r}$$

and the second

$$R = 800\,g = 7848 \text{ N} \qquad \text{as before}$$

Substituting in the first equation for R and for the stated values

$$0.4 \times 7848 = 800 \times \frac{v^2}{40}$$

$$\therefore v = \sqrt{\frac{0.4 \times 7848 \times 40}{800}} = 12.53 \text{ m/s}$$

and if the driver attempted to take the curve faster than this, he would sideslip.

28

Now we shall examine the case of a banked track.

Example A car of mass 800 kg travels at a constant speed of 20 m/s around a circular track of mean radius 40 m. The track surface is banked inwards at an angle of 40°. Evaluate all the forces acting on the car.

The same three forces act as in the diagram of Frame 25, but this time, the friction and the normal reaction will not be horizontal and vertical.

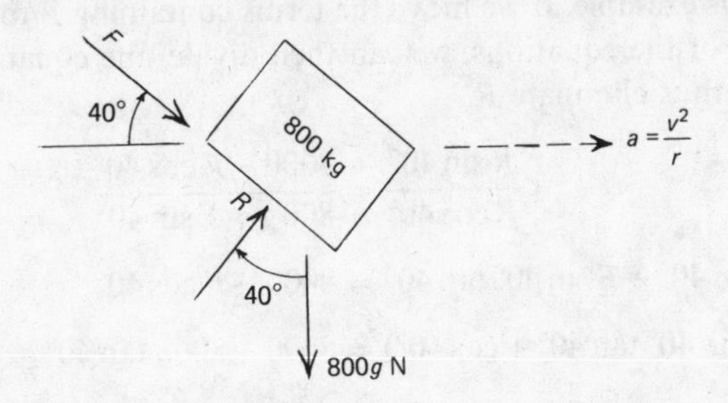

This time, see if you can write the two correct equations. Resolve horizontally and vertically, and not along the track and perpendicular to it. Do you understand why? If not, refer back to Frame 16.

29

Your equations should be

Horizontally,

$$\Sigma F = m \times a: \quad \therefore R \sin 40° + F \cos 40° = 800 \times \frac{20^2}{40} = 8000$$

Vertically, $\Sigma F = 0$: $\quad R \cos 40° - F \sin 40° = 800g$

although it is possible that you may not have them in this exact form.

We require forces F and R. The algebra is more formidable than in the previous example. If we move the terms containing F to the right-hand sides of the equations, we can then divide one equation by the other, and thus eliminate R.

$$\frac{\not{R} \sin 40°}{\not{R} \cos 40°} = \frac{8000 - F \cos 40°}{800g + F \sin 40°}$$

$$\therefore 800g \tan 40° + F \sin 40° \tan 40° = 8000 - F \cos 40°$$

$$\therefore F(\sin 40° \tan 40° + \cos 40°) = 8000 - 800g \tan 40°$$

$$\therefore F = \frac{8000 - 800g \tan 40°}{\sin 40° \tan 40° + \cos 40°}$$

$$= \underline{1083.8 \text{ N}}$$

From the first equation $\quad R = \dfrac{8000 - F \cos 40°}{\sin 40°}$

$$\therefore R = \underline{11\,154.2 \text{ N}}$$

You have probably already realised that we are now able to drive round the bend at a greater speed than the limiting speed of 12.53 m/s without slipping. In fact, for slipping to occur at this particular speed, the value of μ would have to be as low as 0.097 (this is the fraction F/R calculated from the two results above).

Now solve the last part of this example

Example Calculate the greatest speed at which this car may travel around the banked circular track if $\mu = 0.4$.

The two equations will now be

$$R\sin 40^\circ + \mu R\cos 40^\circ = 800\frac{v^2}{r}$$

$$R\cos 40^\circ - \mu R\sin 40^\circ = 800g$$

Dividing: $$\frac{\sin 40^\circ + 0.4\cos 40^\circ}{\cos 40^\circ - 0.4\sin 40^\circ} = \frac{v^2}{rg} = 1.865$$

$$\therefore\ v = \sqrt{1.865 \times 40 \times 9.81}$$

$$= \underline{27.05\,\text{m/s}}$$

When the track is banked, the free-body diagram shows that the centripetal force is provided partly by the frictional force F and partly by a component of the normal reaction R; this is the purpose of banking the track. The vehicle can travel very much faster around a banked track than it can around a flat track, assuming the same friction coefficient. This fact is exploited in the design of racing tracks.

31

'Drill' exercises: dynamics of particles

1. A body of mass 400 kg is located on an inclined plane of slope 35°. The coefficient of friction between body and plane is 0.12. Calculate the acceleration of the body (a) when it moves up the plane, (b) when it moves down the plane.
   Ans. (a) $-6.59\ \mathrm{m\,s^{-2}}$; (b) $+4.66\ \mathrm{m\,s^{-2}}$.
2. A body of mass m is released from rest at the top of an inclined plane of slope 25°. It takes 2.1 seconds to move down a distance of 6 m. Calculate the coefficient of friction between body and plane.
   Ans. 0.16.
3. A wagon ascends a track of slope 10° for a distance of 24 m and then moves on to a horizontal track. Resistance to motion is equivalent to a friction coefficient of 0.04. Calculate the required initial speed of the wagon in order that it will travel 80 m along the horizontal track.

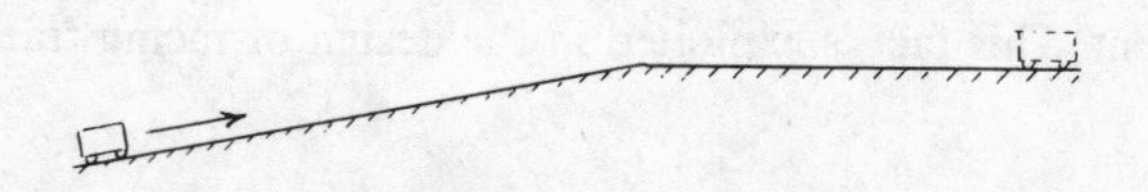

   Ans. 12.77 m/s.
4. A body of mass m rests on the horizontal floor of a truck. μ between body and floor is 0.5. Calculate the least forward acceleration of the truck sufficient to cause the body to begin to slide backwards along the truck floor (a) when the truck travels along a straight level road, (b) when it travels up a straight 15° slope, (c) when it travels down a straight 15° slope.
   Ans. (a) $4.905\ \mathrm{m\,s^{-2}}$. (b) $2.199\ \mathrm{m\,s^{-2}}$. (c) $7.277\ \mathrm{m\,s^{-2}}$.
5. Solve Problem 4 for the case when the floor of the truck is not horizontal, but is tilted backwards at an angle of 10°.

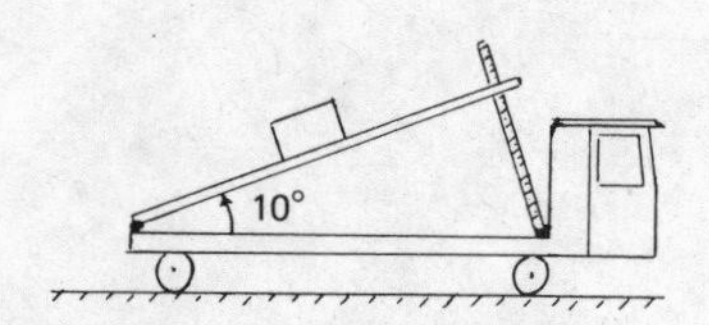

Hint: in all cases, the direction of acceleration of the body will be the same as that of the truck.
Ans. (a) 2.918 m s^{-2}. (b) 0.2795 m s^{-2}. (c) 5.358 m s^{-2}.

6. A small body of mass m at the end of a light cord of length 1.2 m moves in a horizontal circular path, the other end of the cord being attached to a fixed point. (a) If the body completes 1.8 revolutions per second, determine the angle that the string makes with the vertical. (b) Calculate how many revolutions per second the body completes if the string tension is ten times the weight of the body.
Ans. (a) 86.34°. (b) 1.44 rev/s.
7. A locomotive and train has a total mass of 480 Mg. Resistance to motion due to friction is estimated at 1.25 per cent of the total weight. The locomotive exerts a maximum tractive force of 125 kN. Calculate the minimum time for the train to accelerate from rest to a speed of 90 km/hour (a) along a level track, (b) up a slope of $\sin^{-1}$ 0.01, (c) down a slope of $\sin^{-1}$ 0.01.
Ans. (a) 181.4 s. (b) 629.9 s. (c) 106.0 s.
8. The rotor of a small gyroscope is designed to spin at 23,000 rev/min. A small quantity of the rotor metal of mass 1 milligram, at a radius of 25 mm, is accidentally removed. Calculate the resulting unbalanced force exerted on the rotor shaft.
Hint: treat the removed metal as a spinning body of 'negative' mass.
Ans. 0.145 N.
9. A car of mass 780 kg travels at constant speed around a circular track of mean radius 120 m. The track is banked inwards at an angle θ to the horizontal. The coefficient of friction between car and track is 0.46.
   (a) If the car speed is 140 km/hour, calculate the value of θ such that there will be no sideways friction force on the car.
   (b) If angle θ is 40°, calculate the maximum speed at which the car may travel safely.
   (c) If θ is 40° and the car speed is 140 km/hour, calculate the magnitudes of the sideways friction force, and the normal track reaction force, on the car.

   Hint: in (c) friction force is *not* μR; call the force F.
   Ans. (a) 52.1°. (b) 179.7 km/hour. (c) 2611.9 N, 12180.4 N.

32

We now come to the subject of *work*. You may think that you know quite a lot about this subject, having reached this point in this text. But work has a definite meaning in kinetics. It is what is done by an engine, or a machine. In dynamic terms, in order that work can be done, a force has to be brought to bear on an object, and that object must move as a result. The greater the force, the more work done. Also, the more the movement, the more work done. So work is measured as the product of force and displacement. The unit of work (which we could call the newton metre) is actually called the *joule*.

$$\text{Work} = (F \times x)\,\text{joules}$$

The movement must be in the direction of the applied force. If you hold a heavy weight in your hand but do not move, then kinetically you are doing no work. Even if you walk with this weight along a level path, again, you will do no work kinetically, because the movement is horizontal, whereas the applied force is vertical. Only when you begin to go uphill will you be doing actual kinetic work.

Energy is that quality, possessed by a body, or by a substance, or by a system, which enables work to be done. You must be aware that energy exists in many forms. Coal and oil contain energy; a boiler containing high-pressure steam contains energy; the sun contains energy. Less immediately obviously, water in a dam or reservoir is a source of energy, because you can lead the water through pipes and arrange for it to drive turbines, which in turn can produce electrical energy. Moving water is similarly a source of energy, because it can drive a mill or similar machine.

The principal forms of energy of interest to engineers are mechanical, electrical, thermal, nuclear, chemical and solar. Experiment has proved that energy is interchangeable in quantity. By this, we mean that if, for example, we do mechanical work to produce thermal energy, a certain amount of work will always produce exactly the same quantity of energy. The same principle applies in any other form of energy conversion: mechanical to electrical, electrical to thermal, and so on.

The principle of Conservation of Energy has nothing to do with saving energy. It is a principle repeatedly proved by many careful experiments, to the effect that you cannot create or destroy energy; you can only convert it into other forms. For our particular purpose in this text, we can state this principle in the form of an energy equation

Initial energy + Gain of energy − Loss of energy = Final energy

A car travels along a road; it has energy by virtue of its motion. Some energy is added to the system, because fuel is being burned in the car. There is some loss of energy to the system because the car is having to do work against the resistance of the air. If we could accurately measure these quantities, we could then state what the final energy must be. For example, if the energy added were more than the energy lost, the car must go faster.

We must understand one further aspect of energy interchange clearly. Although a specific amount of energy might be available at one point, it is not always possible to convert all of it into another form. For example, we may wish to convert the chemical energy of oil into heat energy in the form of steam in a boiler. But we should be only partially successful, because we could not prevent quite a lot of heat going up the chimney. And if we then try to convert the energy in the steam into mechanical work, we should be lucky to get more than about one-third of it. From this emerges the concept of *efficiency*, which we may define as

$$\text{Efficiency} = \frac{\text{Amount of useful work obtained}}{\text{Amount of energy available}}$$

33

In this text we are concerned only with mechanical energy; this is manifested in two forms—kinetic and potential energy.

Kinetic energy is energy by virtue of the motion of a body. The example of moving water given in the previous frame constitutes kinetic energy. A moving vehicle possesses kinetic energy—because work has been done by a force in order to set it moving. We can derive an expression for this work done, so that we can henceforward calculate kinetic energy, in terms of the mass of a body and its velocity.

Consider a body of mass m acted upon by a constant force F, which acts for a distance x.

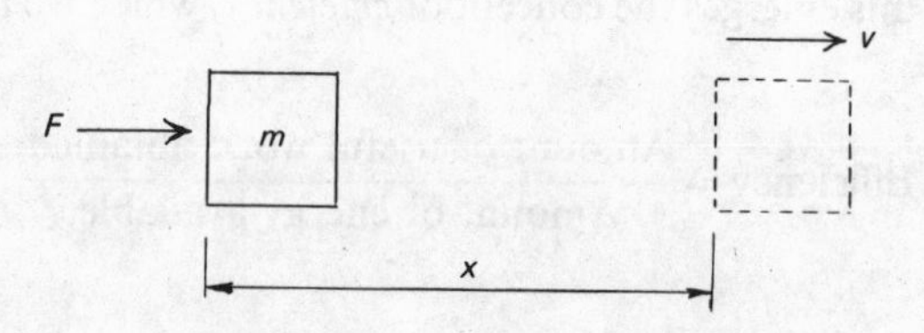

The gain of kinetic energy of the body must be equal to the work done by the force F.

$$\text{Gain of K.E.} = F \times x$$

As a result, the body will be moving with velocity v.

But, from Newton's Second Law: $F = m \times a$

and, kinematically: $v^2 = u^2 + 2ax = 0 + 2ax$

if the body starts from rest.

$$\therefore\ x = \frac{v^2}{2a}$$

$$\therefore\ \text{Gain of K. E.} = (ma)\frac{v^2}{2a} = \tfrac{1}{2}mv^2$$

Potential energy is energy by virtue of the height of a body—above some arbitrary point. The water of a mountain lake can be made to do work, by piping it down to a turbine; the further down you can take it, the more work you can get it to do.

Consider a body of mass m being raised vertically by a force F at constant speed. If the speed is constant, the acceleration will be zero. The free-body diagram is

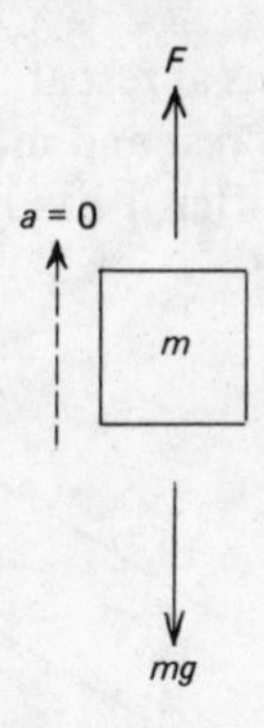

Because acceleration $a = 0$, $F = mg$

Increase of energy = work done by $F = mg \times h$

$\therefore$ Potential energy $= mgh$

In the next few frames we shall use these two formulae to solve simple problems, but just before doing so, it might help for us to rewrite the general equation of energy stated in Frame 32 thus

Initial energy of a body + Work done on body − Work done by body
= Final energy

34

Example A body of mass m is at rest at the top of an inclined track of slope 30°. It is released from rest and moves down the slope without friction. Calculate its speed after it has moved 20 m.

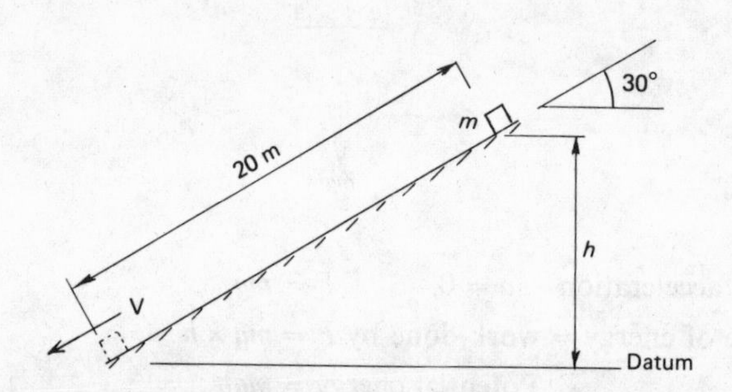

In this example, no work is done *on* the body; it is not being driven, or pulled, or pushed. Neither is any work being done *by* the body; it is not required to overcome friction. So we can say

$$\text{Initial energy} = \text{final energy}$$

Since the body is at rest, initially, the energy can only be potential. At the lower point of the slope, it has lost this potential energy (if we consider the lower point as a 'datum' for estimating potential energy). But it has gained kinetic energy. Thus,

$$\text{Initial (potential) energy} = \text{final (kinetic) energy}$$

$$\therefore mgh = \tfrac{1}{2}mv^2$$

$$\therefore v = \sqrt{2gh}$$

$$= \sqrt{2g \times 20 \sin 30°}$$

$$= \underline{14.0\ \text{m/s}}$$

Let's slightly complicate the example.

Example A body of mass 20 kg is at rest at the top of an incline of slope 30°. It is released from rest and moves down the slope against a friction force of 20 N. Calculate its speed after it has travelled 20 m.

Our energy equation now becomes

Initial potential energy – work done by body = final kinetic energy

and you should be able to arrive at the answer of 12.5 m/s without too much difficulty. Remember: work = force × displacement. The solution follows in Frame 35.

35

$$mg \times h - F \times x = \tfrac{1}{2} mv^2$$

$$\therefore 20\,g\,(20 \sin 30°) - 20 \times 20 = \tfrac{1}{2} \times 20\,v^2$$

$$\therefore v = \sqrt{\frac{(200g - 400) \times 2}{20}} = \underline{12.5\ \mathrm{m/s}}$$

and the body is seen to be going slower than previously; we expect this, as it is being opposed by a friction force.

You need to understand clearly that this is not the 'right' way to solve this problem; it is the simplest way. You could solve both these problems by drawing a free-body diagram, calculating the acceleration, and then using a kinematic equation, and if you think you would gain from this additional practice you should do this for yourself. In doing so, you will find that you have to do more work than we have done here in these last two frames.

36

Here is a different type of example.

> *Example* A small body of mass m hangs vertically at the end of a light string of length $1\frac{1}{2}$ m, the other end being attached to a fixed point. The body is displaced to one side so that the string makes an angle of 60° to the vertical; it is released from rest in this position. Calculate the velocity of the body when it reaches its initial position—that is, when the string is again vertical. Neglect any resistance.

Here, we have no energy added; we are not pushing the body, but releasing it from rest. Similarly, no energy is lost, because no work against friction is to be assumed. The equation is therefore

$$\text{Initial (potential) energy} = \text{final (kinetic) energy}$$

You will require some elementary trigonometry to work out the height to which the body is raised. Take your datum for potential energy through the lowest point. The answer is 3.836 m/s, and the working is shown in the following frame.

37

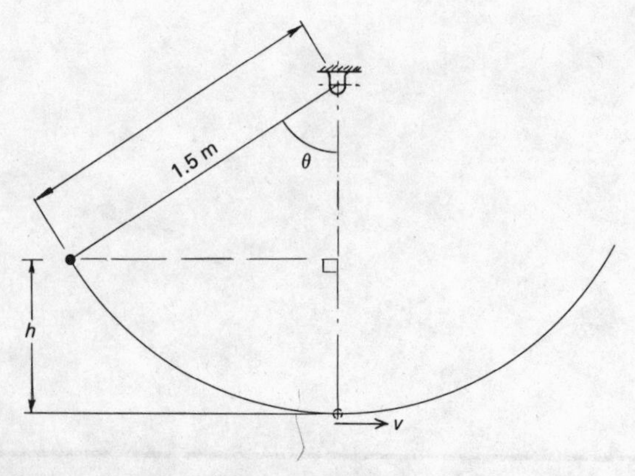

Height: $h = 1.5 - 1.5\cos\theta = 1.5\ (1 - \cos 60°) = 0.75\,\text{m}$

$$\text{P.E.} = \text{K.E.}$$

$$\therefore mgh = \tfrac{1}{2}mv^2$$

$$\therefore v = \sqrt{2gh} = \sqrt{2 \times 9.81 \times 0.75} = \underline{3.836\,\text{m/s}}$$

There are two points of interest in this problem. Firstly, we find that the calculation for v follows exactly the same lines as the example of Frame 34. Moreover, the answer we get is exactly the same as if the body had been allowed to drop vertically from rest through the same height h. In other words, the final velocity of the body is determined only by the vertical height of its fall, and does not depend on how it falls; it may fall vertically, or along a straight slope, or along a curved path. This suggests that an energy approach to a problem is most useful when we are interested only in the states at the beginning and the end of an event, and not in what happens in between.

The second point of interest in this example is that a solution by drawing a free-body diagram and writing an equation of motion is, for our purposes, practically impossible. The difficulty lies in the circumstance that the equation of motion (and therefore the resulting acceleration) is dependent upon the position of the body—that is, the angle of the string. The four equations of kinematics, several of which we have used so far in this programme, are restricted to cases of constant acceleration, and therefore cannot be applied to this case. So this example is typical of a problem which is very easily solved by an energy equation, but which would be extremely difficult to solve by a direct application of Newton's Second Law. But the energy approach can go only so far. It cannot, for instance, tell us *how long* it takes the mass to travel from the highest to the lowest point.

38

Before embarking on the next set of 'Drill' exercises, turn back to Frame 20 and see if you can solve parts (a) and (d) using an energy approach. We'll restate the problem.

> *Example* A body of mass 30 kg is projected up an inclined plane of slope 35° with an initial speed of 12 m/s. The coefficient of friction between body and plane is 0.15. Calculate (a) how far up the slope it travels before coming to rest, (b) how fast it is travelling when it returns to its initial position.

Start with the general equation at the end of Frame 33. The work done on the body will be zero, and the work done by the body will be the product of friction force and distance travelled. You will need to draw a free-body diagram in order to calculate the magnitude of the friction force. You already have the answer. You can check your work against the solution given in Frame 39.

39

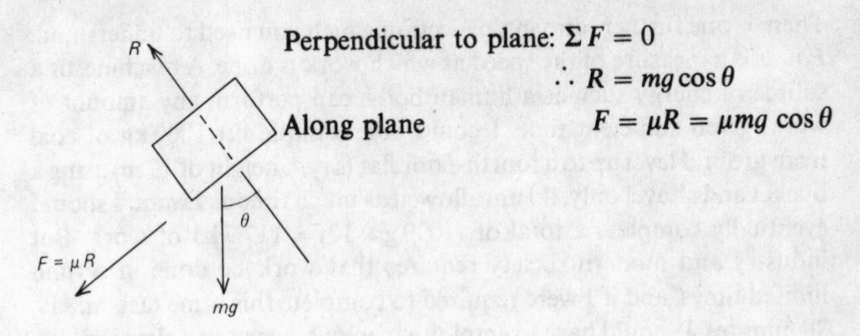

Perpendicular to plane: $\Sigma F = 0$

$$\therefore R = mg\cos\theta$$

Along plane $\qquad F = \mu R = \mu mg\cos\theta$

(a) Initial (kinetic) energy – work done (friction) = final (potential) energy

$$\tfrac{1}{2}mv^2 - (\text{friction}) \times x = mgh$$

$$\tfrac{1}{2}\cancel{m}v^2 - x(\mu \cancel{m}g\cos\theta) = \cancel{m}g(x\sin 35^\circ)$$

$$\tfrac{1}{2} \times 12^2 = x(9.81\sin 35^\circ + 0.15 \times 9.81\cos 35^\circ)$$

$$\therefore x = \frac{72}{9.81(\sin 35^\circ + 0.15\cos 35^\circ)}$$

$$= \underline{10.538\text{ m}}$$

(b) We can cut out the middle stage and work from the initial 12 m/s straight to the final state when the body returns to its starting point. There will thus be no change of potential energy. Simply

Initial (kinetic) energy – work done (friction) = final (kinetic) energy

We have calculated the distance up the plane in (a); the work done against friction will be force × twice this distance, as the friction force opposes motion both up and down.

$$\tfrac{1}{2}\cancel{m} \times 12^2 - 2x(\mu \cancel{m}g\cos\theta) = \tfrac{1}{2}\cancel{m}v^2$$

$$\therefore v^2 = 12^2 - 2 \times 2 \times 10.538 \times 0.15 \times 9.81\cos 35^\circ$$

$$= 144 - 50.809$$

$$= 93.191$$

$$\therefore v = \underline{9.653\text{ m/s}}$$

40

There is one further term in this section which you need to understand. *Power* is a measure of the speed at which work is done. A machine, or a source of energy such as a human body, can perform any amount of work, given sufficient time. I could, for example lift 1000 kg of coal from ground level up to a fourth-floor flat (say, a height of 12 m) using a bucket and shovel only, if I am allowed as much time as I want. I should eventually complete a total of $(1000\,g \times 12) = 117.7$ kJ of work. But industry and modern society requires that work be done in certain limited times, and if I were required to complete this same task in, say, 30 minutes, I should have to employ a machine, such as an elevator, or a crane, capable of an output of 117.7 kJ per 30 minutes, or 65.4 J per second.

The Joule per second is, then, the unit of power, and is given the more convenient name of the *Watt*. Since power is a measure of the speed of work being done, high-power machines are almost always fast machines; a slow machine, however large, is usually a low-power one. The early large steam engines of the Industrial Revolution which are still to be seen in museums have a much lower power rating than a small modern car.

$$\begin{aligned} \text{Power} &= \text{work done per second} \\ &= \text{force} \times \text{displacement per second} \\ &= \text{force} \times \text{velocity} \\ \therefore W &= F \times v \end{aligned}$$

41

Example Calculate the power output of a car of mass 950 kg if it ascends a slope of $\sin^{-1} 0.2$ at a steady speed of 30 m/s against a resistance (due to friction and air) of 3 kN.

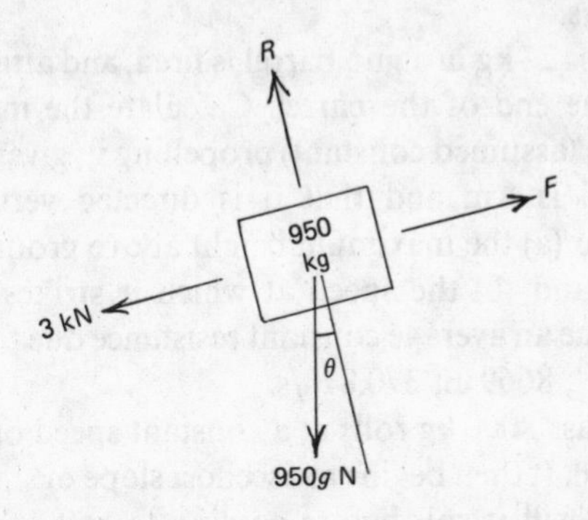

The free-body diagram shows the weight, the normal reaction R, the resistance of 3 kN and the tractive force F. Because the speed is constant, the acceleration is zero.

Along the track: $\Sigma F = 0$: $\therefore F - 3000 - 950g \sin\theta = 0$

$$\therefore F = 3000 + 950 \times 9.81 \times 0.2$$
$$= 4863.9 \text{ N}$$
$$W = F \times v$$
$$= 4863.9 \times 30$$
$$\therefore W = \underline{145.9 \text{ kW}}$$

The resistance force of 3 kN is a mere guess, but it is quite certain that at this speed, a large proportion of the tractive force is used to overcome the resistance due to the air. The search for car bodies having a low drag coefficient takes on more meaning when we look at this simple calculation.

Now tackle the next series of 'Drill' examples in Frame 42.

42

'Drill' exercises: work, energy, power

1. A freely falling body has at one point a downward velocity of 40 m/s. Calculate its velocity after it has fallen a further 40 m. Neglect air resistance.
   Ans. 48.83 m/s.
2. A shell of mass 24 kg in a gun barrel is fired, and attains a velocity of 450 m/s at the end of the barrel. Calculate the magnitude of the average force (assumed constant) propelling it, given that the length of the barrel is 4 m and that it is directed vertically upwards. Calculate also (a) the maximum height above ground level reached by the shell and (b) the speed at which it strikes the ground on return. Assume an average constant resistance due to the air of 45 N.
   Ans. 607.7 kN; 8669 m; 370.8 m/s.
3. A truck of mass 8000 kg rolls at a constant speed of 40 m/s along a horizontal rail. It then begins to ascend a slope of $\sin^{-1} 0.1$. How far up the slope will it roll before coming to rest (a) neglecting any frictional resistance, (b) assuming a resistance due to friction of 1/200th of the weight of the truck?
   Ans. (a) 815.49 m. (b) 776.7 m.
4. Part of a simple 'hump-shunting' railway track comprises a slope of length L at an angle of 15° to the horizontal. Trucks reach the bottom of the slope at a speed of 7 m/s. The track exerts a frictional resistance against the truck of one-eightieth of its weight. Calculate the maximum length L of the track such that the speed of the truck at the top of the slope shall be at least 1.5 m/s.

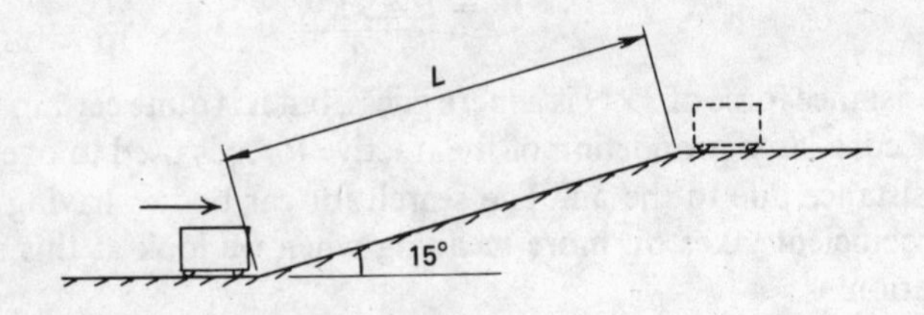

   Ans. 8.78 m.
5. The resistance to motion of a car of mass 850 kg is $(1.2\,v^2)$ newton, where v is its velocity in m/s. Calculate the power output of the car at

speeds of 5, 15 and 30 m/s (a) along a straight horizontal track, (b) ascending a slope of $\sin^{-1} 0.1$. Calculate also the total work done by the tractive force over a distance of 1 km at all three speeds under conditions (a) and (b).

Ans. (a) 150 W; 4.05 kW; 32.4 kW. 30 kJ; 270 kJ; 1080 kJ.
(b) 4.32 kW; 16.56 kW; 57.42 kW. 863.8 kJ; 1103.8 kJ; 1913.8 kJ.

6. A ballistic pendulum comprises a sand-box suspended by four vertical wires of length 1.8 m. A bullet fired into the box is embedded in the sand and causes the box to swing sideways and upwards. The sand-box has a mass of 10 kg and the bullet a mass of 0.1 kg. It may be assumed that only 0.1 per cent of the energy of the bullet is used in moving the box; the rest is dissipated in the sand. If the wires swing through an angle of 18° after impact, estimate the velocity of the bullet just before it strikes the box.

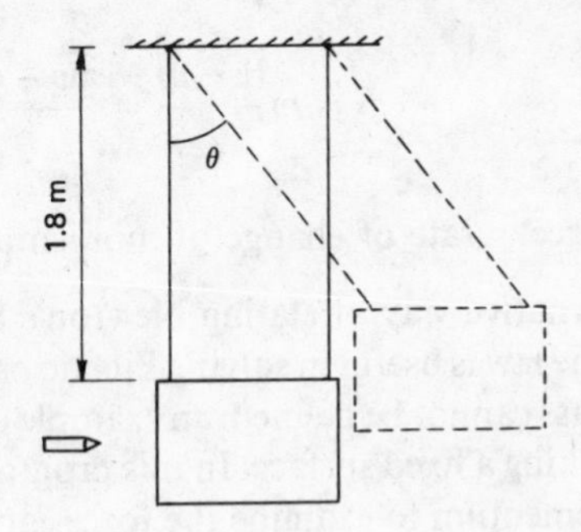

Ans. 417.8 m/s.

7. Water of density 1000 kg/m³ is to be raised by a pump from a reservoir through a vertical height of 12 m at a rate of 5 m³/s. Calculate the required power of the pump. Neglect any energy loss due to friction.
Ans. 588.6 kW.

8. Water emerges from a fire-hose in a jet 50 mm diameter at a speed of 20 m/s. Calculate the power of the pump required to produce this jet, and estimate the vertical height the jet would reach. The density of water is 1000 kg/m³.
Hint: flow rate, kg/s = jet area × velocity × density.
Ans. 7.854 kW; 20.39 m.

43

The term 'momentum' is frequently encountered in kinetics. It is defined simply as (mass × velocity), or (mv). Some textbooks give a qualitative definition of momentum as 'quantity of motion'. Momentum is concerned directly with Newton's Second Law, as we shall now see.

The Second Law states

$$F = m \times a$$

Using the kinematic equation $v = u + at$, we may rewrite this as

$$a = \frac{v - u}{t}$$

$$\therefore F = m\frac{(v - u)}{t} = \frac{mv - mu}{t}$$

In words

Force = rate of change of momentum

and this is an alternative way of stating Newton's Second Law. This interpretation of the law is useful in solving kinetic problems in which a specific value of mass cannot be defined: an example is the force exerted by a jet of fluid striking a fixed surface. In this programme, we shall use this concept of momentum to examine the forces arising when bodies collide. In such a situation, the force of collision is often very large, and the time of contact very short, and both force and time are difficult to determine with accuracy. For this reason, the two quantities are coupled together, and the quantity ($F \times t$) is called the *impulse* of the force. The equation above may be re-arranged

$$(F \times t) = mv - mu$$

In words

Impulse = change of momentum

Let us now consider the case of two bodies, having masses, m_A and m_B, colliding and separating.

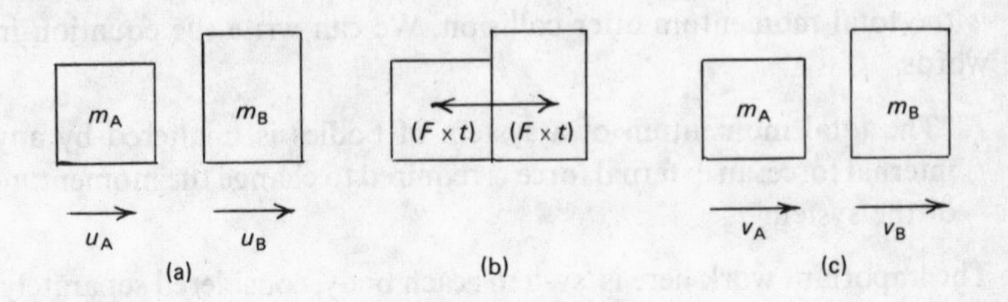

Diagrams (a), (b) and (c) represent before, during and after collision, respectively. Collision will cause an impulse of magnitude $(F \times t)$ to the right on m_B and an impulse of the same magnitude, to the left, on m_A. Because we are considering a general case, we assume that all velocities are from left to right, and this will be our positive direction. Thus, the impulse on m_B will be positive, while that on m_A must be negative. The impulse equations for both bodies are

$$-(F \times t) = m_A v_A - m_A u_A; \; +(F \times t) = m_B v_B - m_B u_B$$

Eliminating $(F \times t)$ from these equations

$$-m_A(v_A - u_A) = +m_B(v_B - u_B)$$

Re-arranging

$$m_A u_A + m_B u_B = m_A v_A + m_B v_B$$

45

The equation obtained at the end of the previous frame is called an equation of Conservation of Momentum: the left-hand side is the total momentum of the system before the collision, while the right-hand side is the total momentum after collision. We can write the equation in words.

> "The total momentum of a system of bodies is unaltered by any internal force: an external force is required to change the momentum of the system"

The important work here is 'system'; each body, considered separately, suffers a change of momentum.

Here is a simple example which you should be able to solve without difficulty.

> *Example* A truck of mass 1200 kg rolls along a straight level track at a constant speed of 4 m/s. It collides with a second truck of mass 1600 kg, which is rolling at $3\frac{1}{2}$ m/s in the opposite direction. The trucks automatically lock together when they collide. Determine the common speed of the trucks after collision.

Watch out: the velocity of the second truck before collision must be negative (if you assume that the velocity of the first is positive). And note also that v_B must equal v_A. The complete solution is given in the next frame, but attempt it yourself before reading on.

46

$$m_A u_A + m_B u_B = m_A v_A + m_B v_B$$

$$1200 \times 4 + 1600(-3\tfrac{1}{2}) = 1200\ v_A + 1600\ v_B = 2800\ v_A$$

$$\therefore v_A = \frac{4800 - 5600}{2800}$$

$$= -\underline{0.2857}\ \text{m/s}$$

The negative sign indicates that the direction of the first truck is reversed when they collide; the second truck continues in its initial direction, but at a reduced speed.

Conservation of Momentum is in one way a somewhat unfortunate phrase, because it makes you think of Conservation of Energy. It might be tempting to try and solve this truck problem by equating the total kinetic energy before collision to the total kinetic energy after collision. A very simple calculation shows that you cannot do this.

$$\text{Total K.E. before collision} = \Sigma(\tfrac{1}{2}\,mu^2) = \tfrac{1}{2} \times 1200 \times 4^2 + \tfrac{1}{2} \times 1600 \times 3\tfrac{1}{2}^2$$

$$= 19{,}400\ \text{J}$$

$$\text{Total K.E. after collision} = \Sigma(\tfrac{1}{2}\,mv^2) = \tfrac{1}{2}(1200 + 1600)\,(0.2857)^2$$

$$= 114.27\ \text{J}$$

You can see that a tremendous amount of energy has been lost to the system. Where has it gone? When bodies collide, there is always a loss of energy because some work is done in deforming the bodies. Sometimes, a deformed body partially or almost completely recovers its original form (a steel spring is a good example) and then, the energy of deformation is restored to the system. But when the deformation is irrecoverable (as, for example, when a football drops into soft mud), a lot of energy is lost to the system. We shall see how to handle this situation in the following frames.

47

Looking back at Frame 44, you can see that the equation of conservation of momentum cannot, by itself, be used to determine the final velocities of bodies after collision; two unknown terms (v_A and v_B) cannot be found from a single equation. In the example we chose, we got round this difficulty by stating that the two final velocities were equal. But in general, some further information is required concerning the nature of the collision.

Imagine a polished steel ball, dropped vertically on to, respectively, a hard steel plate, a wooden table and a bed of plasticine. You can imagine that the height of rebound will vary considerably; in the first instance, it may rebound almost to the height from which it dropped; in the second instance, it certainly would not rebound so high; while in the final instance, it most probably would not rebound at all. In each case, the ratio of the velocity of rebound to the velocity of striking is different. This ratio is given the name *Coefficient of Restitution* and it can be found experimentally for various materials and various types of bodies in collision. It is denoted by *e*. So, by definition, when two bodies collide

$$e = \frac{\text{relative velocity of separating}}{\text{relative velocity of approach}} = \frac{v_A - v_B}{u_A - u_B}$$

and from what we have said, it should be clear that the greatest possible value of e must be 1, and its least possible value 0. There is a very important warning necessary at this point. In the nature of bodies in collision, the velocity of separation must always be in the opposite direction to the approach velocity. (When the ball falls *downwards*, it always rebounds *upwards*.) So, when a value of e is used in problem solving, it must always be given a *negative sign*.

In the following frame, we shall study a problem illustrating the use of this Coefficient of Restitution.

48

Example Two bodies, having masses of 4 kg and 12 kg and travelling along the same straight level path, have respective velocities of 8 m/s right to left, and 2 m/s left to right, when they collide. The coefficient of restitution, e, is -0.75. Calculate the two final velocities and the energy lost due to the collision.

You should be able to start the solution yourself. In order that you can check your attempt against the solution in the following frame, call the 4 kg mass m_A and the 12 kg mass m_B, and assume that a velocity right-to-left is positive. Substituting in the momentum equation will then give you one equation involving v_A and v_B.

49

Using the momentum equation

$$m_A u_A + m_B u_B = m_A v_A + m_B v_B$$
$$4 \times 8 + 12(-2) = 4v_A + 12v_B$$

Simplifying and re-arranging

$$v_A + 3v_B = 2 \qquad (1)$$

and you should have obtained this equation, or some variation on it, yourself.

The remainder of the solution requires the application of the restitution equation

$$e = \frac{v_A - v_B}{u_A - u_B}$$

$$-0.75 = \frac{v_A - v_B}{8-(-2)}$$

Simplifying and re-arranging

$$\left.\begin{aligned} v_A - v_B &= -7.5 \\ v_A + 3v_B &= 2 \end{aligned}\right\} \qquad (2)$$

Rewriting (1): $v_A + 3v_B = 2$

Subtracting: $-4v_B = -9.5$

$$\therefore \underline{v_B = +2.375 \text{ m/s}}$$

Substituting in (1): $v_A = 2 - 3v_B$

$$= 2 - 7.125$$

$$\therefore \underline{v_A = -5.125 \text{ m/s}}$$

Since v_B is positve and v_A negative, it is seen that both masses rebound in the opposite directions after collision.

Loss of kinetic energy $= \frac{1}{2}m_A u_A^2 + \frac{1}{2}m_B u_B^2 - (\frac{1}{2}m_A v_A^2) + \frac{1}{2}m_B v_B^2$

$$= \tfrac{1}{2} \times 4 \times 8^2 + \tfrac{1}{2} \times 12 \times 2^2$$
$$-(\tfrac{1}{2} \times 4 \times 5.125^2 + \tfrac{1}{2} \times 12 \times 2.375^2)$$
$$= 128 + 24 - (52.53 + 33.84)$$
$$= \underline{65.63 \text{ J}}$$

50

Here is a problem of a different kind.

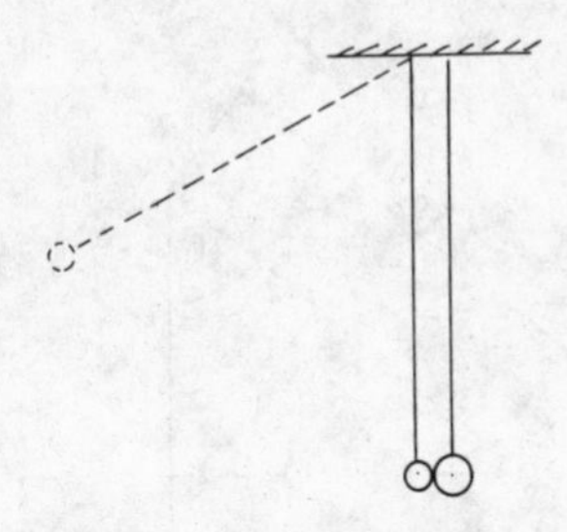

Example Two steel balls of respective masses 4 kg and 12 kg just touch when hanging vertically from strings of length 1.6 m. The smaller one is moved sideways until the string makes an angle of 60° to the vertical; it is then released from rest, and strikes the second ball, the coefficient of restitution of the impact being 1.0. Determine the maximum angle each string makes with the vertical on rebound. Show that there is no loss of energy to the system.

First find the value of u_A (u_B is, of course, 0). You will find it helpful to look back to Frames 36 and 37. You should get a value of 3.962 m/s for u_A. Then make use of the momentum equation and the restitution equation, to find v_A and v_B. Use the values obtained in an energy equation to find the heights of rebound (Frames 36 and 37 again) and thus the angles of the strings. The values of v_B and v_A should be $+1.981$ m/s and -1.981 m/s respectively. The complete solution follows.

51

Equating loss of P.E. to gain of K.E. (see Frames 36, 37)

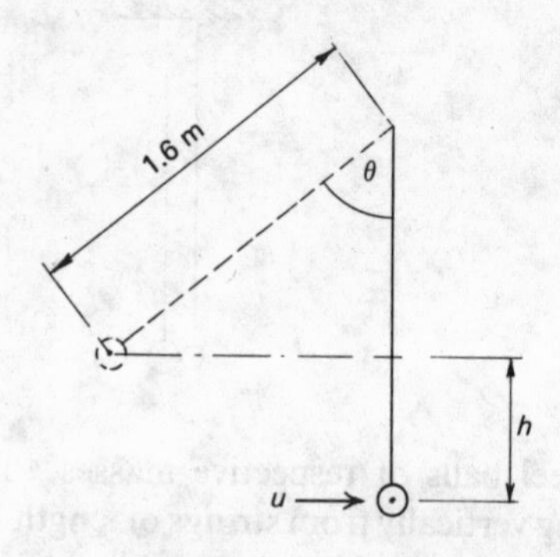

$$mgh = mg \times 1.6(1 - \cos 60°) = \tfrac{1}{2}\, mu_A^2$$

$$\therefore\ u_A = \sqrt{2\, g \times 1.6(1 - \cos 60°)}$$

$$\therefore\ u_A = 3.962 \text{ m/s}$$

Momentum: $m_A u_A + m_B u_B = m_A v_A + m_B v_B$

$$4 \times 3.962 + 12 \times 0 = 4v_A + 12v_B$$

Simplifying $\quad v_A + 3v_B = 3.962 \qquad (1)$

Restitution: $\quad e = \dfrac{v_A - v_B}{u_A - u_B}$

$$-1 = \frac{v_A - v_B}{3.962 - 0}$$

Simplifying: $\quad v_A - v_B = -3.962$

Rewriting (1): $\quad v_A + 3v_B = 3.962 \qquad (2)$

Subtracting: $\quad -4v_B = -7.924$

$$\therefore\ v_B = \underline{+1.981 \text{ m/s}}$$

Substituting in (1): $\quad v_A = 3.962 - 3 \times 1.981$

$$\therefore\ v_A = \underline{-1.981 \text{ m/s}}$$

For the 4 kg ball

Loss of K.E. = gain of P.E.

$$\tfrac{1}{2}m_A v_A^2 = m_A g h_A$$

$$\therefore\ h_A = \frac{v_A^2}{2g} = \frac{(-1.981)^2}{2g} = 0.200 \text{ m}$$

From diagram above: $h_A = 1.6(1 - \cos\theta)$

$$\therefore\ \theta_A = \cos^{-1}\left(1 - \frac{h_A}{1.6}\right) = \cos^{-1}\left(1 - \frac{0.2}{1.6}\right)$$

$$\therefore\ \underline{\theta_A = 28.96°}$$

For the 12 kg ball

$$h_B = \frac{v_B^2}{2g} = \frac{(+1.981)^2}{2g} = 0.200 \text{ m}$$

$$\therefore\ \underline{\theta_B = 28.96°}$$

$$\begin{aligned}\text{Initial P.E.} &= m_A g \times 1.6(1 - \cos 60°)\\ &= 4 \times 9.81 \times 1.6(1 - 0.5)\\ &= 31.392 \text{ J}\end{aligned}$$

$$\begin{aligned}\text{Final P.E.} &= m_A g h_A + m_B g h_B\\ &= 4 \times 9.81 \times 0.200 + 12 \times 9.81 \times 0.200\\ &= 31.392 \text{ J}\end{aligned}$$

A restitution coefficient of 1 means that, although the balls deform under impact, they fully recover their form; in other words, the energy required to deform is fully restored when the bodies regain their original shape. (The word 'restitution' means restoring.) At the other end of the scale, a coefficient of restitution of 0 means that all the energy used in deforming is lost to the system. This does not mean that the final velocities of two masses must both be zero; it simply means that the *relative* velocity of separation after collision is zero— as was the case with the example of Frame 45. In such cases, we do not need to employ the restitution equation; we simply say that $v_A = v_B$.

You ought now to be able to attempt the 'drill' examples in Frame 52.

52

'Drill' exercises: collision of two bodies

1. Two trucks are located on a horizontal frictionless rail. Truck A has a mass of 120 kg and an initial velocity left-to-right of 8 m/s. Truck B has a mass of 180 kg. The trucks collide and lock together automatically on impact. Calculate the common velocity of the two trucks after collision, and also the energy lost to the system, when the initial velocity of Truck B is (a) 6 m/s left-to-right, (b) zero, (c) 6 m/s right-to-left.

   Ans. (a) 6.8 m/s left-to-right, 144 J. (b) 3.2 m/s left-to-right, 2304 J. (c) 0.4 m/s right-to-left, 7056 J.

2. The trucks of Problem 1 collide, with the same velocity conditions stated, but separate after collision, the coefficient of restitution of the impact being 0.8. Calculate the final velocities of the trucks, and the energy lost to the system, for all three cases.

   Ans. (a) 5.84 m/s left-to-right, 7.44 m/s left-to-right, 51.84 J.
   (b) 0.64 m/s right-to-left, 5.76 m/s left-to-right, 829.44 J.
   (c) 7.12 m/s right-to-left, 4.08 m/s left-to-right, 2540.16 J.

3. Two bodies having masses of 40 kg and 80 kg collide when travelling along a straight path. The initial velocity of the smaller body is 5 m/s left-to-right. Determine the initial velocity of the larger body in order that the smaller body rebounds from the collision with an equal and opposite velocity (that is, 5 m/s right-to-left): (a) if the coefficient of restitution of the impact is 1; (b) if the coefficient of restitution of the impact is 0.8. In case (b), calculate the energy lost to the system due to the collision.

   Ans. (a) 2.5 m/s right-to-left. (b) 3.333 m/s right-to-left, 333.33 J.

4. A ballistic pendulum comprises a box filled with sand, hanging from four vertical wires, each 1.5 m long. The mass of the loaded box is 50 kg. A bullet of mass 0.1 kg is fired horizontally into the box. The bullet stays embedded in the sand, and the impact causes the box to swing sideways and upwards, the four wires making a maximum angle of 11° to the vertical. Estimate the bullet velocity before impact.

   Hints: Use an energy equation to determine the velocity of box/bullet immediately after impact (see Frame 51); then use momentum equation to find initial velocity of bullet.

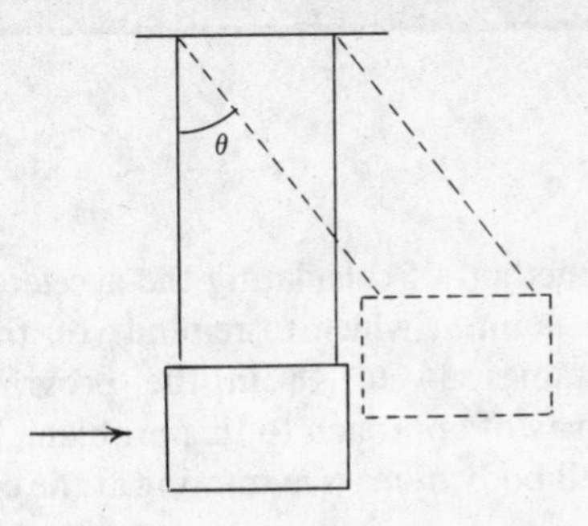

Ans. 368.39 m/s.

5. In a 'Newton's Cradle' apparatus, two steel balls, each of mass 1 kg, hang from vertical wires of length 0.5 m in such a manner that they are just touching when at rest. One ball is moved sideways and upwards, the wire making an angle θ to the vertical, and is then released from rest from this position. Given that the coefficient of restitution of the resulting impact on the other ball is 1, show that the first ball will be brought to rest by the impact, and the second ball projected to the same angle of displacement as the first. If the angle θ is 60° and the time of impact is assumed to be 0.01 seconds, estimate the average force of the impact.
   Ans. 221.5 N.
6. A pile driver comprises a steel block of mass 400 kg. It is dropped from rest through a vertical height of 4 m on to the top of a vertical pile of mass 1500 kg, to drive it into the ground. Assuming that the driver does not rebound on impact, calculate the average resistance of the ground to the pile, given that the blow drives the pile a distance of 44 mm. Estimate the loss of energy to the system resulting from the impact.
   Hints: find common velocity of driver/pile using momentum equation; then write energy equation: initial K.E. + initial P.E. = work done against resistance.
   Ans. 93737 N; 12392 J.

53

Now we need a method of calculating the acceleration of a rotating body. This is the point at which to remind you that you need to be familiar with Frames 34 to 48 in the programme 'Elementary Kinematics'. By way of approach to this problem, look at this simple situation of a small body of mass m rotating at the end of a light rod of radius r

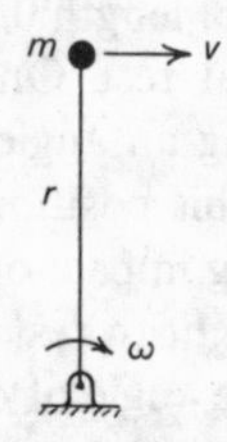

If the angular speed of rotation is ω, the body will have a linear speed v given by $v = \omega r$. The kinetic energy of the system will be $\frac{1}{2}mv^2$. But we need to determine the kinetic energy in terms of the angular speed, not the linear speed of m. Replacing v by (ωr) we get

$$\begin{aligned} \text{K.E.} &= \tfrac{1}{2}m(\omega r)^2 \\ &= \tfrac{1}{2}\omega^2(mr^2) \end{aligned}$$

If instead of a single body of mass m, we have several bodies, all at different radii, we could calculate the kinetic energy of the system in the same way.

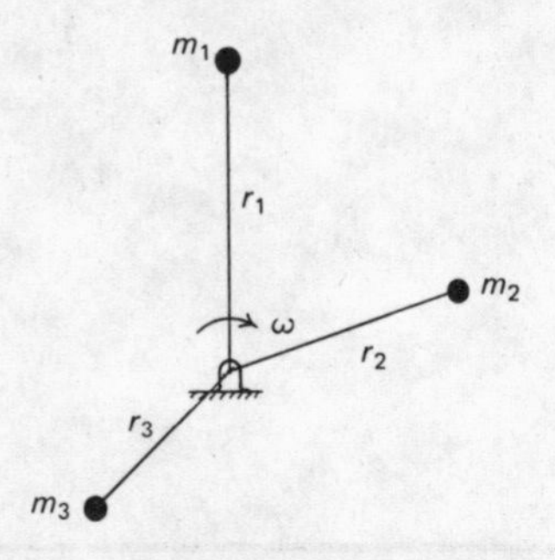

$$\text{K.E.} = \tfrac{1}{2}m_1(\omega r_1)^2 + \tfrac{1}{2}m_2(\omega r_2)^2 + \tfrac{1}{2}m_3(\omega r_3)^2$$
$$= \tfrac{1}{2}\omega^2(m_1 r_1^2 + m_2 r_2^2 + m_3 r_3^2)$$

We can express this more simply as

$$\text{K.E.} = \tfrac{1}{2}\omega^2(\Sigma(mr^2))$$

The bracketed quantity $(\Sigma(mr^2))$ is called the *Moment of inertia* of the system, and it is usually denoted by the letter I. So we can say even more briefly

$$\text{K.E.} = \tfrac{1}{2}I\omega^2$$

If you compare this expression with the K.E. of a mass m having velocity v

$$\text{K.E.} = \tfrac{1}{2}mv^2$$

you see that the moment of inertia I replaces the mass m, while angular velocity replaces the linear velocity v. We begin to see I as a sort of 'rotational mass' and indeed, this is not a bad way to think about moment of inertia, provided that you realise from the outset that it does not have the dimensions of mass. You may recall in Frame 2 of this program that we treated mass as 'resistance to acceleration'. We might therefore expect to see moment of inertia as 'resistance to angular acceleration', and this is exactly what we shall find in the following frame.

54

Look back at the simple single body system at the beginning of Frame 53. To increase its speed we would have to apply a force. But unlike a simple body travelling along a straight line, the increase of speed would depend not only on how much force we applied, but also on where we apply it. Imagine a large wheel, say 3 or 4 metres diameter, mounted on a horizontal shaft. Suppose you have the task of turning the wheel round. You might attempt to twist the shaft with your hands, or you might apply a force at the rim of the wheel. I do not think you will need actually to try this to appreciate that the wheel would move much more readily in response to the force at the rim than to one at the shaft, even if the two forces were equal in magnitude. It is the *moment* of the force which changes the motion, rather than just the force itself. How do we estimate the work done by the moment of a force? Consider a force F which has a moment M about some point O, and let the force move through a linear distance x.

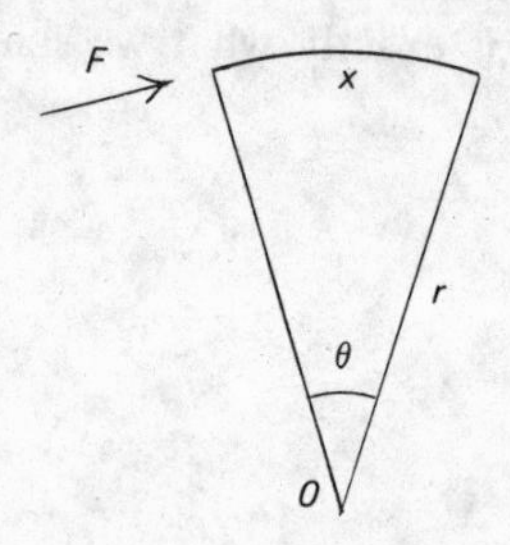

The work done by the force $= F \times x$. But the moment M of the force about O is $F \times r$

$$\therefore \text{work done} = \left(\frac{M}{r}\right)x$$

But in angular measure, $x = \theta \times r$

$$\therefore \text{work done} = \left(\frac{M}{r}\right)(\theta r) = M\theta$$

Compare this with the expression for linear work done

$$\text{work done} = F \times x$$

and the expressions are seen as analogous, in that work = moment × displacement in the same way that work = force × displacement for linear work (although the dimensions of the analogous terms are not the same).

So now imagine a moment M applied to our simple 1-mass system, and let it displace the system through an angle θ. As a result, let the angular speed increase from ω_1 to ω_2.

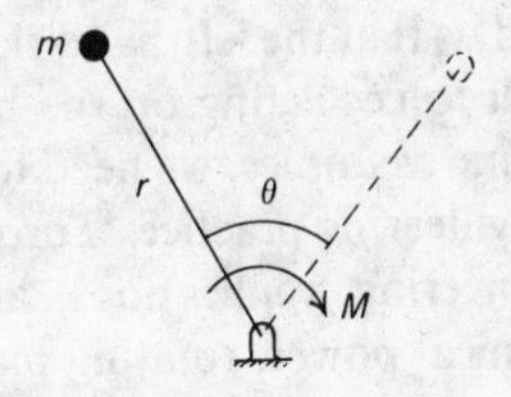

$$\text{Work done} = \text{increase of kinetic energy}$$

$$\therefore\ M\theta = \tfrac{1}{2}I(\omega_2^2 - \omega_1^2)$$

But from angular kinematics we know that $\omega_2^2 = \omega_1^2 + 2\alpha\theta$

$$\therefore\ M\theta = \tfrac{1}{2}I(2\alpha\theta)$$

$$\therefore M = I \times \alpha$$

Compare this with $F = m \times a$ and again, we find the terms are analogous, and that we are justified in regarding the moment of inertia I as "resistance to angular acceleration."

So the fundamental equation of Newton's Second Law applied to rotating bodies is

$$\Sigma M = I \times \alpha$$

(The Σ serves to remind us that the left-hand side of the equation is the *resultant* moment of all forces acting on the body).

Before looking at some examples, we need to explain another word which is common in engineering practice. 'Torque' is another word for 'Twist'. In many engineering applications, and particularly in the transmission of mechanical power, rotating machinery often forms a useful way of conveying energy from one point to another. In an internal combustion engine such as is used in an automobile, the combustion of the fuel is caused to exert a force on a moving piston; this force is then caused, by the mechanism of a connecting-rod and a crank, to apply a twist, to *torque*, to the engine shaft. When the engine is used, to drive a vehicle, or a generator, or a pump, the person using it is not directly concerned with the detailed mechanism; it is sufficient that he knows the magnitude of the torque that the shaft is capable of producing (together, of course, with the speed). Similarly, when an engineer uses an electric motor to drive a machine, he does not need to know the magnitude of the electro-magnetic force between the rotating armature and the fixed stator; the torque and speed of the motor shaft are sufficient. Because torque is produced by a force applied in such a manner as to turn a shaft, it is measured in the same units as moment, and in the examples that follow, you can think of torque and moment as synonymous.

It is interesting to observe that some modern developments have been successful in breaking through what might be termed the 'torque barrier'. When aircraft became practically possible, it was considered obvious that they should be driven by a rotating propeller. This was almost certainly a direct consequence of the influence of the automobile and the petrol engine; indeed, several early designs used modified car engines. But the jet engine by-passes this mode of transmission, which now appears as a clumsy and archaic method of propelling aircraft. We can observe a corresponding development in electric traction in the 'linear motor' in which the tractive force on a vehicle is the direct electro-magnetic repulsion force between an armature and a 'flat' stator of indefinite length, fixed to the ground.

We can now look at a simple example.

Example A wheel, initially at rest, is subjected to a torque of magnitude 250 N m (newton-metres) and attains a speed of 2200 rev/min in 8 seconds. Calculate the moment of inertia of the wheel.

We can determine α from the kinematic equation

$$\omega_2 = \omega_1 + \alpha \times t$$

$$2\pi \times \frac{2000}{60} = 0 + \alpha \times 8$$

$$\therefore \alpha = \frac{2\pi \times 2000}{60 \times 8}$$

$$= 26.18 \text{ rad s}^{-2}$$

$$\therefore I = \frac{M}{\alpha} = \frac{250}{26.18} = \underline{9.549 \text{ kg m}^2}$$

56

Now you do one.

Example A wheel of moment of inertia 12 kg m² is turning at 850 rev/min. A braking torque reduces the speed to 200 rev/min in 5.5 seconds. Calculate the magnitude of the torque.

Your answer should be 148.5 N m. Check your work in the following frame.

57

$$\omega_2 = \omega_1 + \alpha t$$

$$\frac{2\pi \times 200}{60} = \frac{2\pi \times 850}{60} + \alpha \times 5.5$$

$$\therefore \alpha = -\frac{2\pi(850 - 200)}{60 \times 5.5} = -12.376 \text{ rad s}^{-2}$$

$$M = I\alpha = -12 \times 12.376$$

$$= \underline{-148.5 \text{ N m}}$$

The negative sign is consistent with the information that the torque is a braking (that is, retarding) torque.

58

How are we to calculate moments of inertia? What is the mathematical procedure for evaluating the quantity $\Sigma(mr^2)$? Consider a development of the three-body system of Frame 53; we shall extend this to an infinite number of infinitely small bodies all rotating about a fixed centre, that is, a continuous solid rotating body.

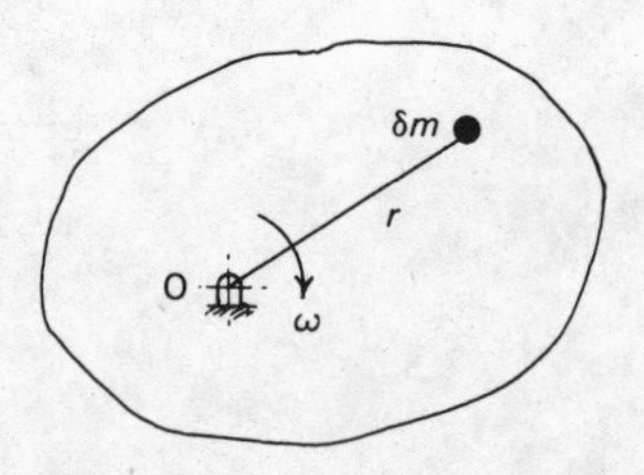

If we examine a very small 'element' of the body, of mass δm at a radius r (assuming the element to be small enough to be considered as all at the one radius), our original definition of moment of inertia I can be modified to,

$$I = \Sigma(\delta m \times r^2)$$

and henceforward, this more general expression should be taken as a standard definition of moment of inertia.

One fairly simple method of evaluating I would be to divide a body into a number of actual elements. Provided that we make them small enough to be considered as being concentrated at one single radius, without too much obvious error, we could then 'sum up' all such elements to evaluate I. An example should make this clear.

Example The diagram shows a half-section of a uniform steel wheel of outer and inner radii 0.2 m and 0.1 m, and width 0.3 m. The density of the steel is 7800 kg/m³. Evaluate the moment of inertia of the wheel by dividing into five rings, as shown, each of radial thickness 0.02 m, considering each 'element' as acting at its mean radius.

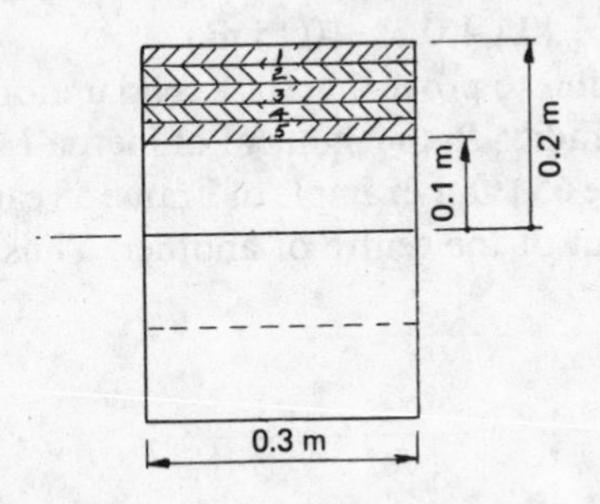

For the outermost ring, the volume is $0.3 \times \pi(0.2^2 - 0.18^2)$

$$\therefore \ \delta m_1 = 0.3\pi(0.2^2 - 0.18^2) \times 7800 = 55.87 \text{ kg}$$

The mean radius is $\frac{1}{2}(0.2 + 0.18) = 0.19$ m

$$\therefore \ \delta m_1 r_1^2 = 55.87 \times 0.19^2 = 2.017 \text{ kg m}^2$$

The remaining calculations are not given in detail but are tabulated below.

Element	*Inner radius* (m)	*Outer radius* (m)	δm (kg)	*Mean radius* (m)	$\delta m \times r^2$ (kg m²)
1	0.18	0.2	55.87	0.19	2.017
2	0.16	0.18	49.99	0.17	1.445
3	0.14	0.16	44.11	0.15	0.992
4	0.12	0.14	38.23	0.13	0.646
5	0.10	0.12	32.35	0.11	0.391
			$\Sigma(\delta m)$ 220.55 kg		$\Sigma(\delta mr^2)$ 5.491 kg m

59

This procedure of dividing a body into 'elements' is, of course, tedious, but may occasionally be necessary when a wheel has an irregular shape of rim cross-section. When the cross-section is rectangular, as in the example, we can use calculus to derive a formula, and such a formula will then give us an accurate value for moment of inertia. You need to understand that the tabular calculation of the previous frame is not absolutely accurate, because assuming each element to have an effective radius equal to its mean radius is actually an approximation. (If it were not so, we could assume the whole wheel to be a single 'element' at an effective radius of $\frac{1}{2}$ (0.1 + 0.2) = 0.15 m.)

We can use calculus to prove that for a solid uniform wheel or disc of mass m and outer radius R, the moment of inertia $I = \frac{1}{2}mr^2$. (We shall prove this in Frame 65.) Our example of Frame 58 can be treated as one solid disc taken out of the centre of another. Thus

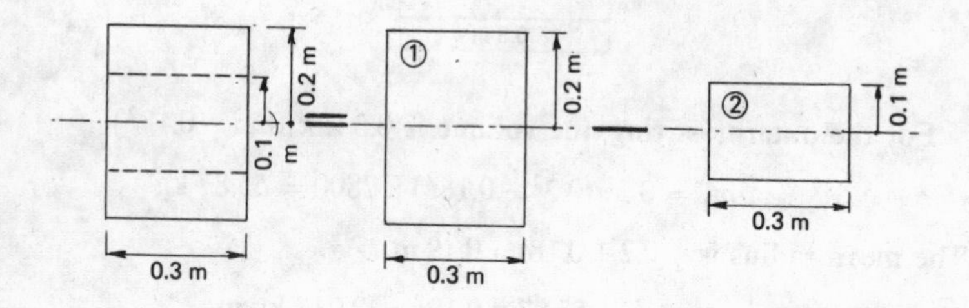

So

$$I = I_1 - I_2$$

and since both I_1 and I_2 are solid discs, we can use the formula $I = \frac{1}{2}mr^2$ to calculate them. Assuming the same value for density of 7800 kg/m

$$\begin{aligned} I &= \tfrac{1}{2}(7800 \times 0.3 \times \pi \times 0.2^2)0.2^2 - \tfrac{1}{2}(7800 \times 0.3 \times \pi \times 0.1^2)0.1^2 \\ &= \tfrac{1}{2} \times 7800 \times 0.3\pi(0.2^4 - 0.1^4) \\ &= \underline{5.513 \text{ kg m}^2} \end{aligned}$$

and our answer using the division into 5 'elements' is seen to be about 0.4 per cent lower than the correct value.

Quite complex wheels can be analysed by treating them as assemblies of solid uniform discs. See if you can 'break down' the wheel shown in the section-drawing below, into a system of simple discs.

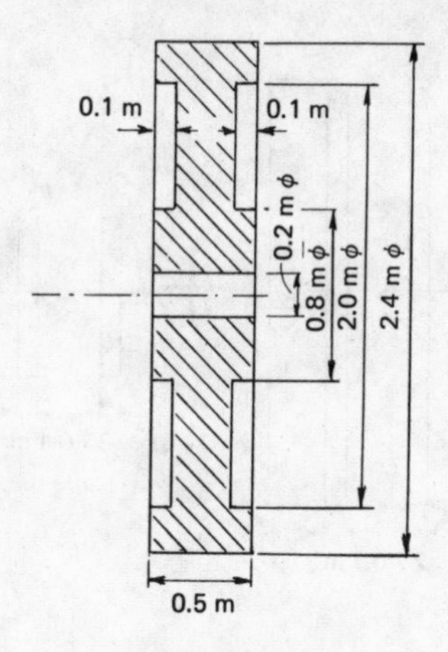

60

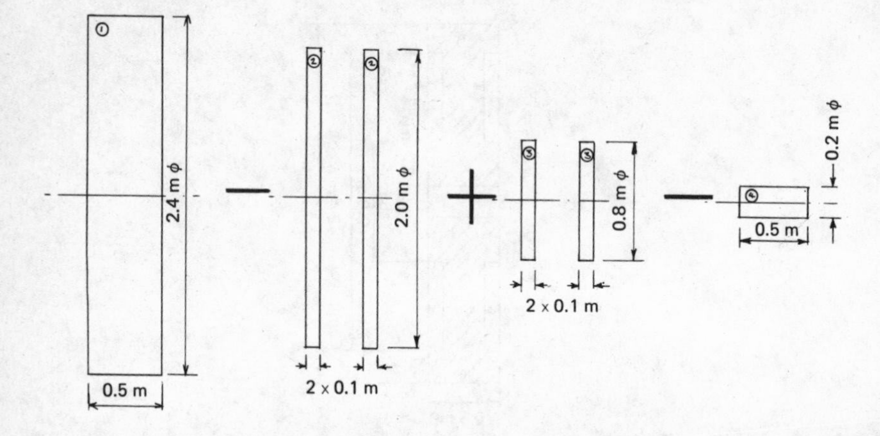

Elements ① and ③ are 'positive' elements while ② and ④ are 'negative'. Although there are really six elements making up this wheel, you can see that there are two pairs, so that we may calculate on the basis of four different elements.

Taking the density of the wheel material as 7800 kg/m^3 as before, and writing

$$I = I_1 - I_2 + I_3 - I_4$$

calculate the mass and the moment of inertia of each of the four elements, and hence calculate I for the whole wheel of the previous frame. The total mass is 13,404 kg, and the total moment of inertia is 10,314.8 kg m^2, (work to one place of decimals). If you fail to get these values, check carefully against the working shown in Frame 61.

$$
\begin{aligned}
m_1 &= \quad 7800 \times 0.5 \times \pi \times 1.2^2 && = \quad 17643.2\,\text{kg:} \\
-m_2 &= -7800 \times 0.2 \times \pi \times 1.0^2 && = -4900.8\,\text{kg:} \\
m_3 &= \quad 7800 \times 0.2 \times \pi \times 0.4^2 && = \quad 784.1\,\text{kg:} \\
-m_4 &= -7800 \times 0.5 \times \pi \times 0.1^2 && = -122.5\,\text{kg:} \\
&\text{Total mass} && = \quad 13404.0\,\text{kg}
\end{aligned}
$$

$$
\begin{aligned}
I_1 &= \tfrac{1}{2} m_1 R_1^2 = \tfrac{1}{2} \times 17643.2 \times 1.2^2 && = \quad 12703.1\,\text{kg m}^2 \\
I_2 &= \tfrac{1}{2} m_2 R_2^2 = \tfrac{1}{2} (-4900.8)\, 1.0^2 && = -2450.4\,\text{kg m}^2 \\
I_3 &= \tfrac{1}{2} m_3 R_3^2 = \tfrac{1}{2} \times 784.1 \times 0.4^2 && = \quad 62.7\,\text{kg m}^2 \\
I_4 &= \tfrac{1}{2} m_4 R_4^2 = \tfrac{1}{2} \quad (-122.5) 0.1^2 && = \quad -0.6\,\text{kg m}^2 \\
&\text{Total moment of inertia} && = 10314.8\,\text{kg m}^2
\end{aligned}
$$

These two columns of results repay some study. Notice, for example that while the mass of element 4 is roughly one-sixth of that of element 3, its moment of inertia is only about one-hundredth. This illustrates the principle that in order to have a large moment of inertia, a mass must be distributed at as large a radius as is possible. A flywheel is a special type of wheel, designed to absorb a large amount of energy. It is attached to the shaft of an engine (such as a car engine) to keep the speed fairly steady, when the driving torque is unsteady. Without a flywheel, a single-cylinder petrol engine would surge to a high speed throughout the driving stroke, and drop to inconveniently low speed during the three non-driving strokes. The flywheel keeps the speed steadier. Such a wheel has as much as possible of its mass distributed as far as possible from the axis of rotation.

62

The only really accurate way of calculating moment of inertia is by the use of calculus (and even this is of no use if the wheel has an irregular shape). But we can often make a shrewd guess at the value of I. For example, if we have a thin ring rotating about its axis, say of outer radius 0.2 m and inner radius 0.18 m, it seems reasonable to assume that, dynamically, this would behave as though all the mass were at the mean radius of 0.19 m. (This is actually what we did in Frame 58, with the result that the calculated value of I was accurate to within 0.4 per cent). But the thicker the ring, the less accurate it becomes to assume a mean radius. You can see why this should be so by looking at a thick 'thin' ring

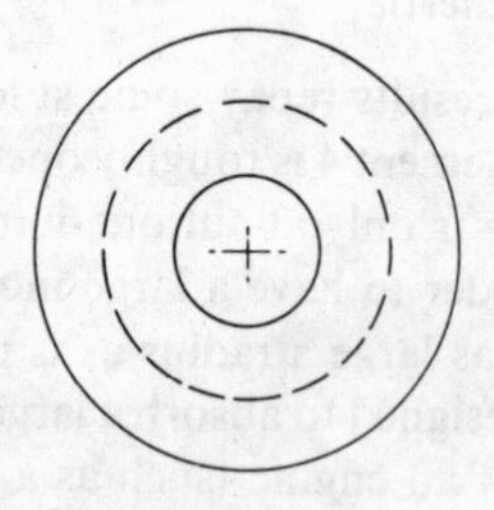

It is clear that there is much more material on the outside of the mean radius than on the inside; assuming a mean radius will therefore give a value for the moment of inertia which is too low. (The answer of 5.491 kg m calculated in Frame 58, although close to the correct value, is actually low, not high.) The thicker the ring, the greater the error in assuming a mean radius to calculate I. You can check for yourself that if we treat the ring of Frame 58 as a single 'element' at a mean radius of 0.15 m, the resulting value of I would be 4.962 kg m: this is approximately 10 per cent lower than the correct value. Staying for a moment with this particular example, it is again easy to show, by working backwards, that if we had assumed an 'effective' radius of 0.158 m instead of the mean radius of 0.15 m, we should get an accurate value for the moment of inertia. This 'effective' radius—the radius at which all the mass would have to be in order to yield the same value for moment of inertia as the actual wheel—is called the radius of gyration, and it is usually denoted by the letter k.

Assuming all the mass to be concentrated at a single radius k, we can say

$$I = mk^2$$

Use this formula to calculate k for the wheel shown in Frame 58, using the results in the table at the end of that frame.

63

$$I = mk^2$$

$$\therefore k = \sqrt{\frac{I}{m}} = \sqrt{\frac{5.491}{220.55}} = \underline{0.158 \text{ m}}$$

This is the value we suggested must be used in the previous frame, in order to calculate the moment of inertia accurately.

Now look back to Frame 59, at the wheel illustrated at the end of the frame. The value of moment of inertia I was calculated for this wheel in Frame 61. Use the result to calculate k for this wheel. Your answer should be 0.877 m.

64

$$k = \sqrt{\frac{I}{m}} = \sqrt{\frac{10314.8}{13404.0}} = \underline{0.877 \text{ m}}$$

This is slightly over two-thirds of the outer radius of the wheel.

It is quite common to specify the radius of gyration of a wheel instead of its moment of inertia. This has an advantage from a practical point of view. It is not difficult to determine the mass of a wheel, either by calculating it, or by weighing the wheel. And it is possible to estimate the value of radius of gyration with reasonable accuracy—say, within 20 per cent of the correct figure, just by looking at the section of a wheel. Thus, the moment of inertia can be estimated, easily and quickly, if not absolutely accurately. Of course, if a really accurate value is required, a guess or estimate is not sufficient, and an accurate calculation must be made (using the calculus, or the method of Frame 58) or I must be determined from a dynamic experiment.

For the student, the concept of radius of gyration is also useful in that it constitutes a useful check on calculated values of moment of inertia. For example, in Frame 59, we calculated a value of 5.513 kg m^2 for I for the wheel shown in Frame 58. We know (from Frame 58) that the mass is 220.55 kg. Thus

$$k = \sqrt{\frac{I}{m}} = \sqrt{\frac{5.513}{220.54}} = 0.158 \text{ m}$$

which is somewhere about the mean radius, and as such, is likely to be right. So the corresponding value of I is also likely to be right. But suppose we had made a mistake in calculation, and obtained a value for I of 55.13 kg m^2 instead of 5.513 kg m^2. This would yield a value for k of 0.5 m and this could not be correct: the 'effective' radius must clearly lie somewhere between the inner and outer boundaries of the wheel.

To conclude this section, we shall now determine the moment of inertia of a solid uniform disc, using calculus.

65

Consider a thin incremental ring element as shown, of inner radius r and thickness δr. Let the material density be ρ, and the width of the wheel be B.

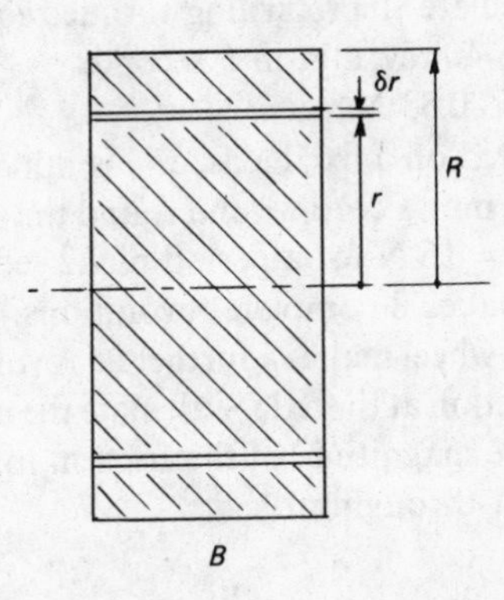

The mass of the element, $\delta m = \rho \times 2\pi r \times \delta r \times B$

$$I = \Sigma\,(\delta m \times r^2) = \Sigma\,((\rho \times 2\pi r \times \delta r \times B)r^2)$$

$$= \rho \times 2\pi B \int_0^R (r^3)\mathrm{d}r$$

$$= \rho \times 2\pi B \left[\tfrac{1}{4}r^4\right]_0^R$$

$$= \rho \times 2\pi B \times \tfrac{1}{4} R^4$$

Total mass of wheel, $m = \rho \times \pi R^2 B$

$$\therefore\ I = \tfrac{1}{2} m R^2$$

Now work through the 'drill' examples on moment of inertia in Frame 66 before continuing with the text.

66

'Drill' exercises: moment of inertia; rotating bodies

1. A wheel has a moment of inertia of 120 kg m^2. Calculate the torque required to accelerate the wheel to a speed of 80 rev/min in 20 revolutions. Calculate the retarding torque required to reduce the speed from 80 to 40 rev/min in 5 seconds.
   Ans. 33.51 N m; 100.53 N m.
2. A wheel is mounted on a fixed axis, and is initially at rest. The axle has a radius of 28 mm; a cord wound round the axle is subjected to a constant force $F = 15$ N for a period of 7.2 seconds, during which time the wheel makes 8 complete revolutions. The force F is then removed, and the wheel makes a further 25 revolutions in coming to rest due to the friction at the axle. Calculate the moment of inertia of the wheel, and the magnitude of the friction torque, which may be assumed constant throughout.

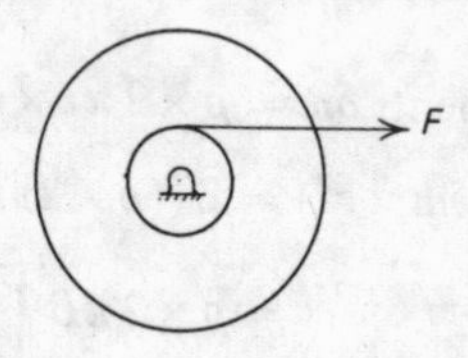

   Ans. 0.164 kg m^2; 0.1018 N m.
3. The cross-section of a cylindrical wheel is shown. Calculate the moment of inertia of the wheel, and its radius of gyration, given that $B = 0.3$ *m*, $R_1 = 0.8$ m, and $R_2 =$ (a) 0.1 m; (b) 0.3 m; (c) 0.75 m. Use the formula $I = \frac{1}{2}mR^2$ which applies to a solid uniform cylinder, and assume the material to have a density of 7800 kg/m. For each case, calculate the percentage error introduced by assuming a radius of gyration k equal to the mean radius, $\frac{1}{2}(R_1 + R_2)$.

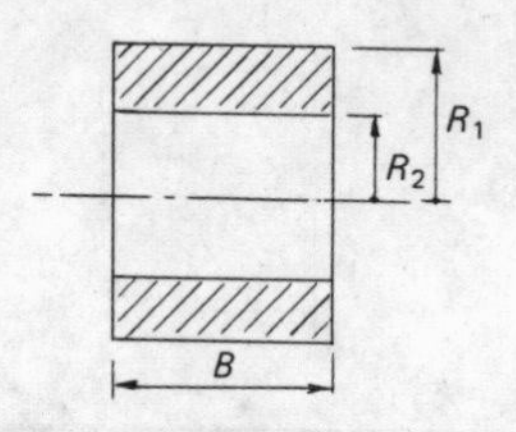

Ans. (*a*) 1505.18 kg m^2; $k = 0.57$ m; -37.7 per cent. (b) 1475.77 kg m^2; 0.604 m; -17.1 per cent. (c) 342.55 kg m^2; 0.775 m; 0.103 per cent.

4. Using only the formula $I = \frac{1}{2} mR^2$ for the moment of inertia of a solid uniform cylinder about its axis, evaluate the moments of inertia of the four wheels shown in cross-section below, and their radii of gyration. Assume a density of 7800 kg per cubic metre for the wheel material in each case.

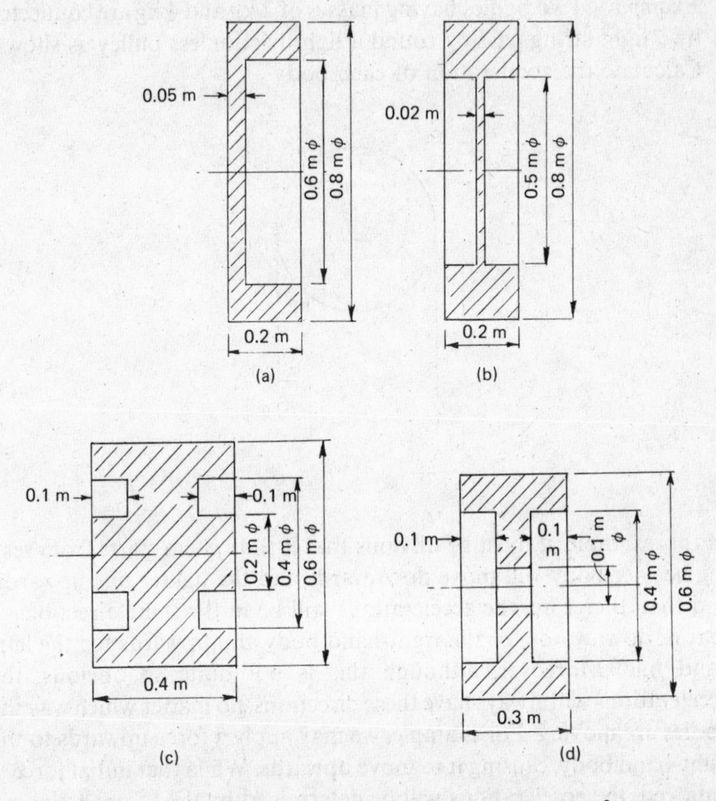

Ans. (a) 47.84 kg m^2; 0.325 m. (b) 54.12 kg m^2; 0.326 m. (c) 36.02 kg m^2; 0.221 m. (d) 25.84 kg m^2; 0.237 m.

5. A flywheel for an engine is to be designed to fluctuate ± 2 per cent about a mean speed of 280 rev/min and the corresponding fluctuation of energy is required to be 450 J. Calculate the required moment of inertia of the wheel. If it is to be an annular ring, of outer radius R, inner radius $0.8\,R$, and width $0.2\,R$, of material of density 7600 kg/m^3, calculate the value of R.
Ans. 13.09 kg m^2; 0.392 m.

67

In this section, you have to learn to analyse systems of bodies in which one body is connected to another in some way. Before laying down a procedure, we can attempt some simple examples.

Example Two bodies having masses of 2 kg and 4 kg are connected by a light string passing round a light frictionless pulley as shown. Calculate the acceleration of each body.

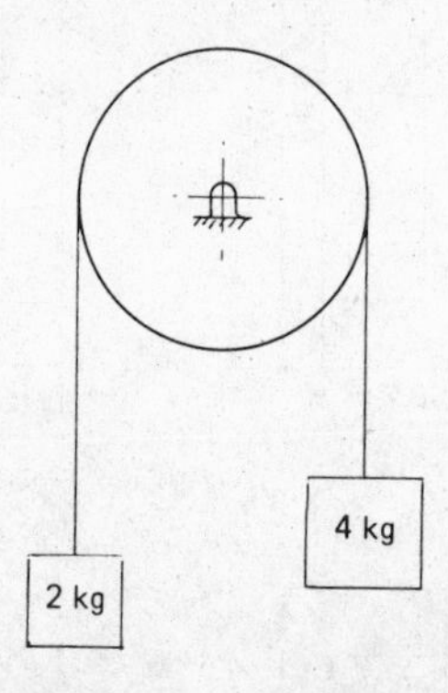

In this example, it must be obvious that if the system starts from rest, the heavier body will move downwards and the lighter one upwards, and that, therefore, the accelerations will be in the same directions—that is, downwards for the right-hand body and upwards for the left-hand one. Moreover, although this is not quite so obvious, the accelerations will always have these directions, no matter which way the bodies are moving. For example, we may apply a force upwards to the right-hand body, causing it to move upwards. When that initial force is removed, the accelerations will be determined by the forces acting on the system shown, and will thus be downwards for the right-hand body and upwards for the other. The right-hand body will therefore continue moving upwards, but with a reducing velocity; it will eventually come to rest, and will then begin to move downwards. One lesson to be learned from this situation is, not to confuse motion with acceleration.

We require equations of motion for the two bodies. At this stage, refresh your memory in respect of free-body diagrams; look back to Frame 9 and the checklist. You should then be able to see that each body is subjected to two forces—weight and string tension. Weight acts downwards in both cases, and string tension acts upwards in both cases (remember—a string cannot push). A common error at this stage, when considering the forces on the 4 kg mass, is to assume that the string tension is the weight of the 2 kg mass. If this were so, then the forces acting on the 2 kg mass would be (2 g) N downwards (weight) and (2 g) N upwards (tension). Thus there would be no resultant force, and therefore it could not accelerate. From this argument, we can draw two conclusions

(a) the tension is less than the weight of the 4 kg mass
(b) it is more than the weight of the 2 kg mass.

For the present, we shall resist the temptation to assume that it is (3 g) N (half-way between both weights) and simply call it T. We have assumed, by the way, that it is the same tension on both ends of the string; this follows from the facts that it is a single string, and that the wheel is light and frictionless. We shall later be looking at cases wherein the tension varies in a single string.

So, now draw the two free-body diagrams, showing on each one the arrow for the acceleration. Do not forget the earlier suggestion to use a red pen for force arrows, to distinguish from accelerations. Also write the two equations of motion.

68

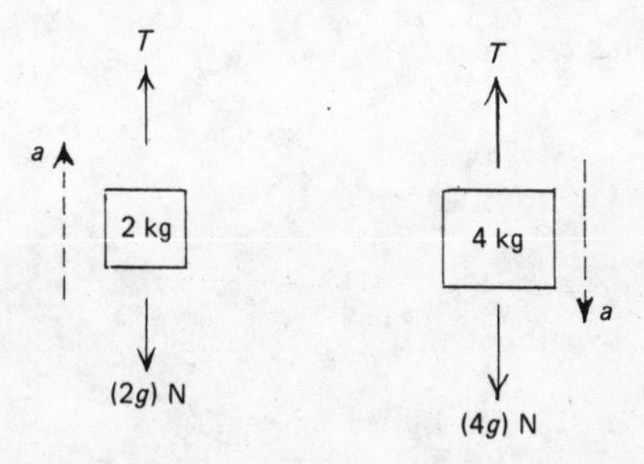

With the free-body diagrams correctly drawn, the writing of the two equations of motion becomes simple: that, of course, is the purpose of drawing the diagrams. So write the two equations, and solve the problem by eliminating T from them. And although the question does not ask for it, when you have calculated a, substitute back in either of the equations, and find T. You should obtain a value for a of 3.27 m s^{-2} and T should be 26.16 N. The working follows in the next frame.

69

For the 2 kg mass: $T - 2g = 2a$ (1)

For the 4 kg mass: $4g - T = 4a$ (2)

Adding: $4g - 2g = 6a$

$$a = \frac{2 \times 9.81}{6}$$

$$= \underline{3.27 \text{ m s}^{-2}}$$

Re-arranging (1): $T = 2a + 2g$

$$= 2(3.27 + 9.81)$$

$$= \underline{26.16 \text{ N}}$$

70

Simple? Perhaps so. If you didn't arrive at the correct answers, the important thing is that you find out what you did wrong. Now try this next problem from the beginning.

Example

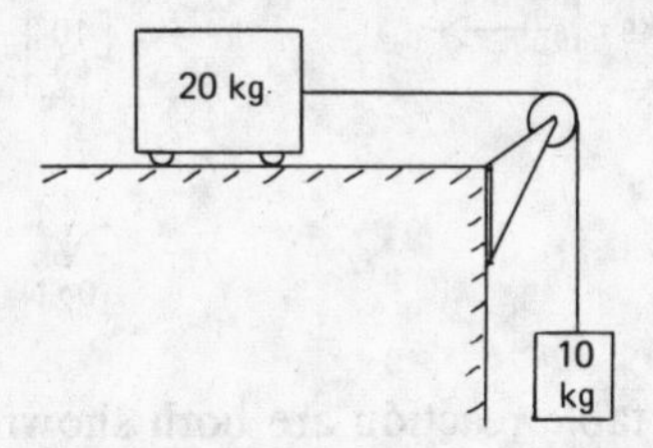

A body of mass 20 kg rests on a smooth horizontal table. It is connected by a light string to a body of mass 10 kg which hangs freely, the string passing over a light frictionless pulley. Determine the acceleration of the bodies, and the string tension.

Conditions are similar to the first problem; the acceleration is the same for both bodies, although not in the same direction. The string tension is the same. Have a go and check your work in the next frame.

71

Here are the free-body diagrams

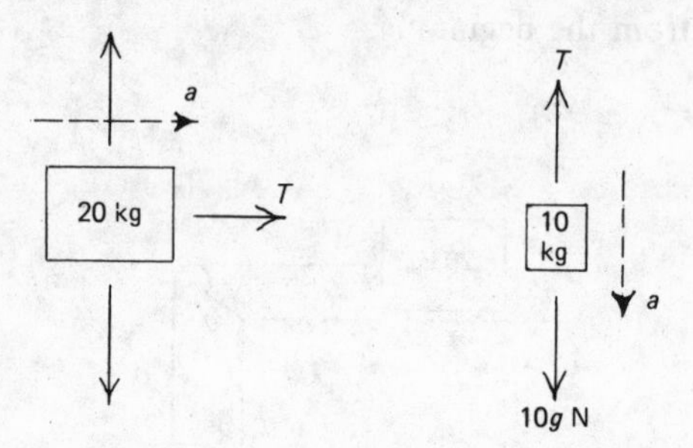

The weight and the table reaction are both shown on the free-body diagram for the larger body, because they are two of the forces acting, but we have not bothered to include values or letters, because we shall not need these forces to solve the problem. The two equations are

For the 20 kg

$$T = 20a \tag{1}$$

For the 10 kg

$$10g - T = 10a \tag{2}$$

Adding:

$$10g = 30a$$

$$\therefore\ a = \tfrac{1}{3}g = \underline{3.27 \text{ m s}^{-2}}$$

Substituting in (1)

$$T = 20 \times 3.27 = \underline{65.4 \text{ N}}$$

In the two examples so far, the directions of the accelerations have been (we hope) obvious. But sometimes one cannot immediately decide, just by casual inspection, which way a body will accelerate (in other words, which will be the direction of the resultant force acting on it). Look at this next example.

Example Determine the accelerations of the two bodies shown on the inclined planes, and the tension in the connecting string. Neglect all friction forces.

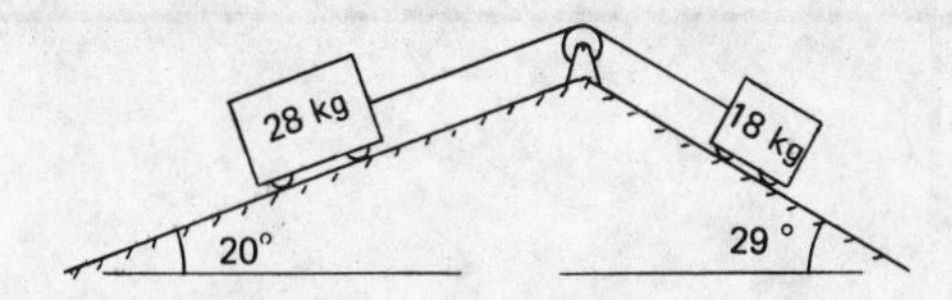

In Frames 13 to 15 we solved a problem in which the acceleration was assumed to be in one direction and was shown (by a negative value) to be in the other. We can do the same with this problem, with one further reservation. It should be clear that the line of action of acceleration must be along the slope in each case. We may assume the acceleration of the right-hand body to be acting down the slope. If we are wrong, we shall get a negative value. *But*, having assumed this direction, we *must* assume that the acceleration of the other body will be *up* the slope: the two accelerations must be, as we say, *compatible*. There is no way, short of the string breaking, whereby the accelerations of both bodies can be down their respective slopes.

With this proviso, go ahead and draw the diagrams, and write the equations. The string tension (T) will be constant throughout.

72

The diagrams

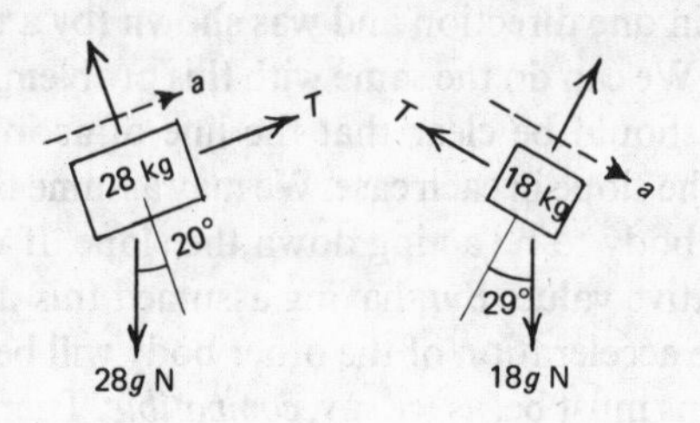

The equations

For the 28 kg: $T - 28g \sin 20° = 28a$ (1)

For the 18 kg: $18g \sin 29° - T = 18a$ (2)

Adding: $18g \sin 29° - 28g \sin 20° = 46a$

Re-arranging: $a = \frac{9.81}{46}(18 \sin 29° - 28 \sin 20°)$

$$= \frac{9.81}{46}(8.727 - 9.577)$$

$$a = \underline{-0.181 \text{ m s}^{-2}}$$

Substitute in (1): $T = 28(-0.181) + 28 \times 9.81 \times 0.342$

$$\therefore T = \underline{88.87 \text{ N}}$$

The negative sign attached to a means that the acceleration is down the incline for the 28 kg mass and up for the 18 kg mass.

73

The next example introduces friction forces. When friction is involved, care is necessary in order to determine the direction of the friction force in each case. Friction always acts in the opposite direction to the motion of a body; thus, if the direction of motion changes, the force system, and therefore the acceleration, changes. In any problem involving friction forces, therefore, the direction of motion must be clearly stated, or clearly implied.

Example

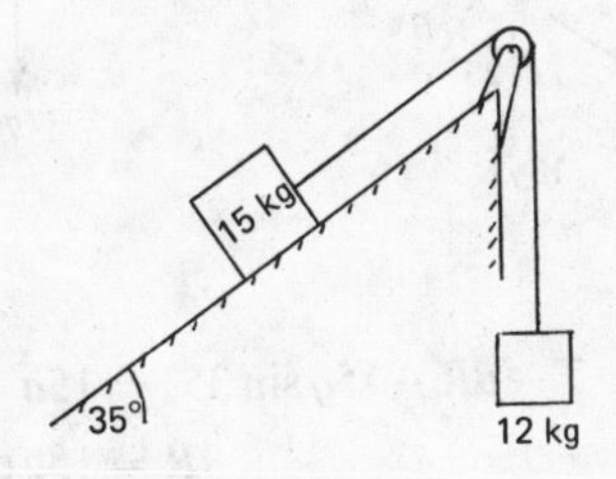

The 15 kg body rests on a plane inclined at 35° as shown. The coefficient of friction between body and plane is 0.24. A light string connects the body to a second body of mass 12 kg which is hanging freely. If the 15 kg body is moving up the plane, calculate its acceleration.

Draw the two free-body diagrams; assume the acceleration to be directed up the plane for the 15 kg body. Then write the two equations, as before. This time, you will find that you haven't enough information to find a; you will need a third equation—an equation of force equilibrium transverse to the 15 kg mass. Check your work in the following frame.

74

Here are the two free-body diagrams. Although the direction of acceleration is not actually known (we have only assumed it) the direction of the friction force is in no doubt—since the body moves up the plane, this force must act down it.

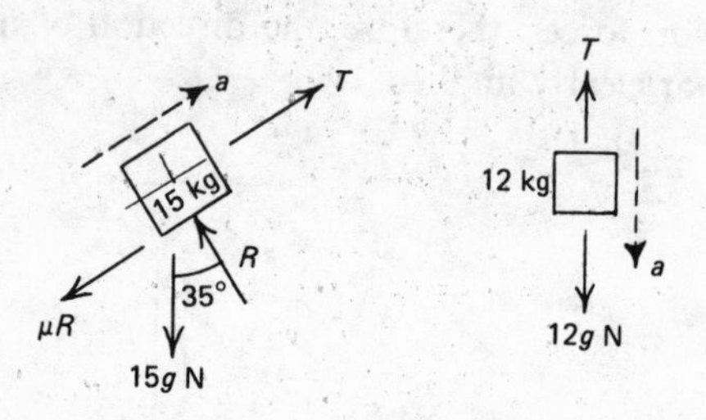

The equations

Along the plane: $T - \mu R - 15g \sin 35° = 15a$ (1)

Across the plane: $R = 15g \cos 35°$ (2)

Hanging body: $12g - T = 12a$ (3)

Adding (1) and (3) and substituting for R in (1) at the same time

$$12g - 0.24(15g \cos 35°) - 15g \sin 35° = 27a$$

$$\therefore a = \frac{9.81}{27}(12 - 0.24(15 \cos 35°) - 15 \sin 35°)$$

$$= \underline{0.1625 \text{ m s}^{-2}}$$

If you would like to give yourself a little more practice, find the acceleration of this same system, but this time with the 15 kg body moving down the plane, not up. The solution is not given, but the answer you should get is 2.305 m s^{-2} up the plane.

75

Example

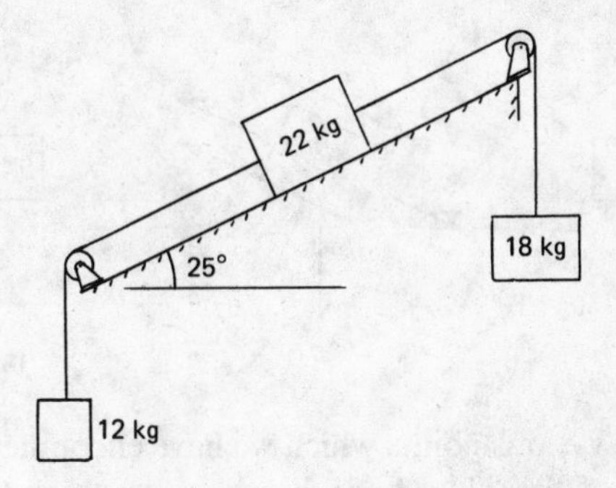

For the system of three bodies shown, the coefficient of friction between the 22 kg body and the plane is 0.32, and the body is moving down the plane. Determine the accelerations of the bodies, and the tensions in the strings.

Start this one yourself. So that you can check your work, assume that the 22 kg body is accelerating down the plane. There are *two* separate strings in this problem, so you will have two different string tensions, T_1 and T_2. You must start by drawing the three free-body diagrams, and writing four equations. (You will need two equations for the 22 kg body. Check your diagrams in Frame 76 before proceeding.

76

Here are the diagrams

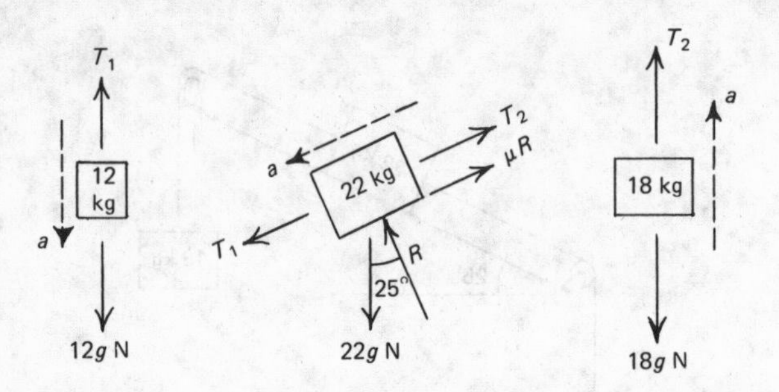

Note again the various points which we have encountered before. The acceleration a of the 22 kg body has been assumed to be down the plane; therefore, that of the 12 kg body, to be compatible, must be vertically downwards, and similarly, that of the 18 kg body must be vertically upwards. These directions may be wrong (we shall see by the later working that they are) but this will not matter provided that they are all compatible—that is, consistent with one another. Note also that string tension forces can only pull a body, they cannot push.

Now write the equations, recalling that the resultant force on each body must be considered positive in the (assumed) direction of the acceleration.

77

For 12 kg body: $$12g - T_1 = 12a \qquad (1)$$

For 22 kg body; along plane:

$$T_1 + 22g \sin 25° - T_2 - \mu R = 22a \qquad (2)$$

and across plane: $$R = 22g \cos 25° \qquad (3)$$

For 18 kg body: $$T_2 - 18g = 18a \qquad (4)$$

The rest is algebra. Find T_1 from (1) and T_2 from (4) and R from (3) and substitute appropriately in (2). The complete working follows.

78

Substituting in equation (2):

$$(12g - 12a) + 22g \sin 25° - (18a + 18g) - \mu \times 22g \cos 25° = 22a$$

Re-arranging

$$a(22 + 12 + 18) = 12g + 22g \sin 25° - 18g - \mu \times 22g \cos 25°$$

$$\therefore \ a = \frac{9.81}{52}(12 + 22 \sin 25° - 18 - 0.32 \times 22 \cos 25°)$$

$$\therefore \ a = \underline{-0.582 \text{ m s}^{-2}}$$

Substitute in (1): $T_1 = 12g - 12a = 12 \times 9.81 - 12(-0.582) = \underline{124.7 \text{ N}}$
Substitute in (4):

$$T_2 = 18a + 18g = 18(-0.582) + 18 \times 9.81 = \underline{166.1\ N}$$

79

How do we proceed when one of the bodies of a system is a rotating wheel? For an introduction to this type of problem, solve the following example. If you experience difficulty, go back and look at Frames 53 to 66 and do a bit of revision.

Example

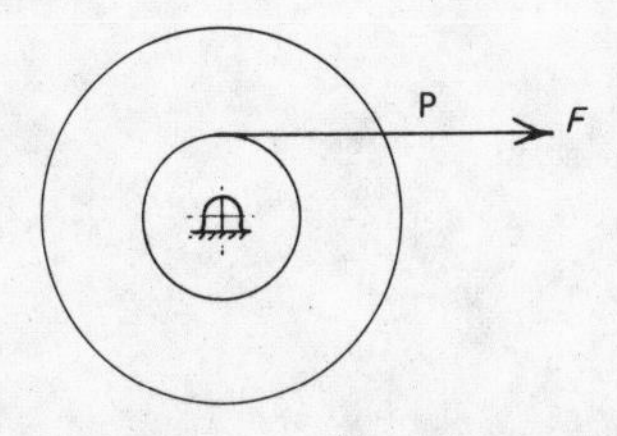

A wheel rotating about a fixed axis has a moment of inertia of 49 kg m². A rope is wrapped around the axle of radius 75 mm and a force F = 72 N applied to the rope. Calculate the angular acceleration of the wheel.

80

For rotating bodies $\Sigma M = I \times \alpha$

$$\therefore \alpha = \frac{\Sigma M}{I} = \frac{(72 \times 75 \times 10^{-3})}{49}$$

$$= 0.1102 \text{ rad s}^{-2}$$

Not too difficult. While on this particular example, see if you can now calculate the *linear* acceleration of a point P on the rope (perhaps you need to brush up your elementary kinematics).

81

Relating linear to angular acceleration

$$a = \alpha \times r = 0.1102 \times 75 \times 10^{-3} = 0.008265 \text{ m s}^{-2}$$

With this little bit of revision dealt with, we can now have a look at a typical simple problem of connected bodies, one of which is rotating.

Example

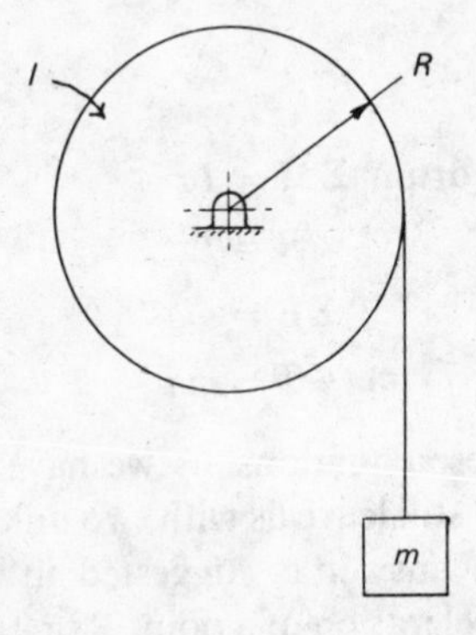

A body of mass $m = 3.5$ kg hangs by a light cord which is wound round a drum of moment of inertia $I = 1.6 \text{ kg m}^2$ at an effective radius $R = 0.24$ m. Calculate the angular acceleration of the drum and the linear acceleration of the mass.

Draw free-body diagrams for both the drum and the hanging body. Be careful *not* to assume that the tension in the cord is equal to the weight of the body. On the free-body diagram for the drum, you should actually find three forces acting, but two of these (the weight, and the upward bearing reaction force) will have the line of action passing through the centre of the wheel. In determining $\Sigma(M)$, the resultant moment on the drum, only the cord tension has a moment about the drum centre. Derive your two equations of motion and check in the next frame.

82

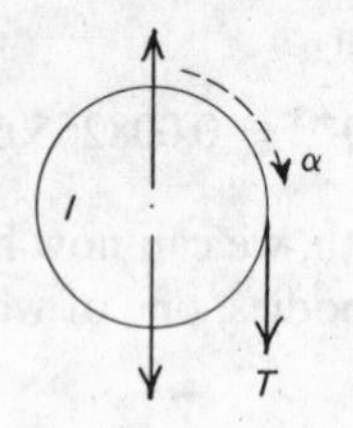

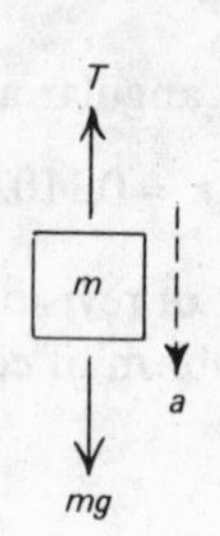

The equations: for the drum: $\Sigma M = I\alpha$

$$\therefore T \times R = I\alpha \quad (1)$$

for the hanging body: $\Sigma F = ma$

$$\therefore mg - T = ma \quad (2)$$

We can manipulate these equations, as we have previously done, to eliminate T, but this will still leave us with two unknown terms, α and a. What can we do? The answer is suggested in Frame 80 when we calculated the linear acceleration of a point P on the rope; we can write a third equation relating a and α. Thus

Using elementary kinematics: $a = \alpha R$ (3)

From equation (2): $T = mg - ma$

Substitute in equation (1):

$$(mg - ma)R = I\alpha$$

Substitute for 'a': $(mg - m \times \alpha R)R = I\alpha$

Re-arranging: $mgR = I\alpha + m\alpha R^2$

$$\therefore \alpha = \frac{mgR}{I + mR^2}$$

$$= \frac{3.5 \times 9.81 \times 0.24}{(1.6 + 3.5 \times 0.24^2)}$$

$$\therefore \alpha = \underline{4.574 \text{ rad s}^{-2}}$$

Substitute in (3): $a = \alpha R$

$$= 4.574 \times 0.24$$

$$\therefore a = \underline{1.098 \text{ m s}^{-2}}$$

Some rules for solution are now beginning to appear.

1. Draw free-body diagrams for each separate body.
2. Assume directions for all accelerations, linear and angular, but *check that these are all compatible.*
3. Write appropriate equations of motion ($\Sigma F = ma, \Sigma M = I\alpha$) for all bodies, taking the (assumed) direction of acceleration as positive in every case.
4. Where necessary, write equations of equilibrium for bodies in directions transverse to the direction of acceleration.
5. Write kinematic equations of compatibility relating all accelerations.
6. Check that you have sufficient equations (as many equations as there are unknown terms) and then manipulate algebraically.

Refer back to these rules when attempting the example in the next frame.

83

Example

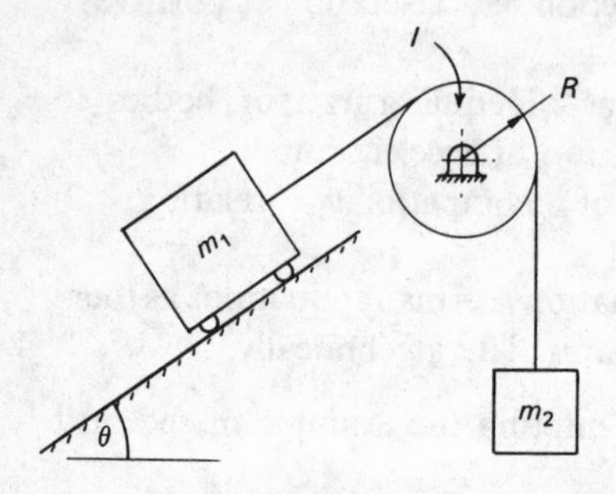

$m_1 = 12$ kg
$m_2 = 5$ kg
$I = 1.4$ kg m
$R = 0.15$ m
$\theta = 34°$

Bodies of masses m_1 and m_2 are connected by a light cord which is wrapped around the drum which rotates about a fixed axis. The rope does not slip on the drum. Friction between m_1 and the inclined plane is negligible and there is no friction at the drum axis. Determine the linear acceleration of the two bodies, and the angular acceleration of the drum.

A word of warning. The direction of acceleration of the bodies is not immediately obvious. So assume m_1 to accelerate up the plane. You must therefore assume the angular acceleration of the drum to be clockwise; an anti-clockwise acceleration is incompatible with the first assumption. The direction of acceleration of m_2 must, clearly, be downwards.

Another warning before you begin. Although the two bodies are connected by a single cord, you *must not* now assume that the tension will be the same throughout the cord, as we have so far done. We are told that the rope does not slip on the drum; therefore it must either be secured to it, or it must grip it by friction. In either case, this will cause a difference of tension between the rope ends either side of the drum. When you draw the free-body diagram of the drum, this becomes even more obvious. If the tensions were the same, there could be no resultant torque on the drum, and therefore it could not accelerate. So call the tensions T_1 and T_2 appropriately. You may find that your work is more systematic and orderly if you retain algebraic symbols throughout, and substitute numbers only at the end. This is what is done in the solution that follows.

84

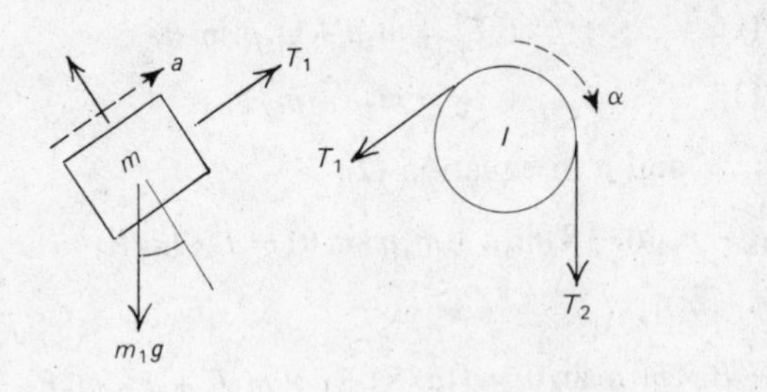

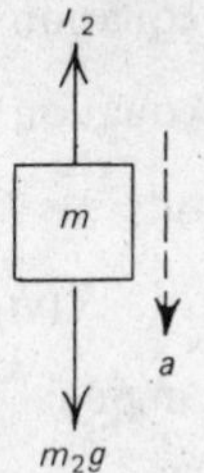

For m_1	$T_1 - m_1 g \sin\theta = m_1 a$	(1)
For drum:	$T_2 R - T_1 R = I\alpha$	(2)
For m_2:	$m_2 g - T_2 = m_2 a$	(3)
Kinematically:	$a = \alpha R$	(4)

You may have found by now that you can obtain the correct equations without too must difficulty, but that you run into trouble afterwards, when you try to manipulate the equations. The more equations you have, the more the chance of trouble. Unfortunately, there are no firm rules to go by for this part of the work, and there is no substitute for experience in this type of work. In this present problem, equation (2) contains both T_1 and T_2, and you can re-arrange equation (1) to make T_1 the subject, and equation (3) to make T_2 the subject. You can then substitute for T_1 and T_2 in equation (2). But only lots of practice will make you competent in this sort of calculation.

85

From equation (1): $T_1 = m_1 a + m_1 g \sin\theta$

From equation (3): $T_2 = m_2 g - m_2 a$

Substitute for T_1, T_2 and α in equation (2):

$$R(m_2 g - m_2 a) - R(m_1 a + m_1 g \sin\theta) = I \times (a/R)$$

Re-arranging:

$$R \times m_2 g - R \times m_1 g \sin\theta = I(a/R) + a \times m_1 R + a \times m_2 R$$

$$\therefore\ a = \frac{m_2 g - m_1 g \sin\theta}{I/R^2 + m_1 + m_2}$$

$$= \frac{9.81(5 - 12 \sin 34^\circ)}{\dfrac{1.4}{0.15^2} + 12 + 5}$$

$$\therefore\ \underline{a = -0.2118\ \mathrm{m\,s^{-2}}}$$

the negative sign indicating that all accelerations are in the opposite directions to those assumed.

From equation (4): $\alpha = \dfrac{a}{R} = \dfrac{-0.2118}{0.15}$

$$\therefore\ \underline{\alpha = -1.412\ \mathrm{rad\,s^{-2}}}$$

86

One last example before you embark on the revision examples.

Example

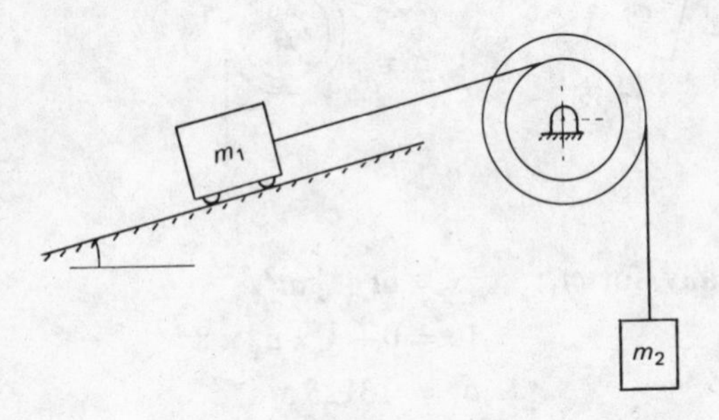

A truck of mass $m_1 = 1250$ kg is hauled up a slope of 15° by a simple hoist which consists of a compound drum, of radii 0.24 m and 0.32 m, total moment of inertia 250 kg m^2, and a counter-weight of mass m_2, 200 kg. The hauling rope is attached to the inner radius and the counter-weight to the outer, as shown. The hoist is operated by a motor (not shown) driving the axis of the drum. Calculate the magnitude of the driving torque required at the drum axis in order to pull the truck up the slope a distance of 10 m from rest in 8 seconds. Assume a resistance to motion of the truck of 800 N.

Use your elementary kinematics first to calculate the linear acceleration of the truck. Then draw the free-body diagrams. And do not forget to include, on the drum diagram, the motor driving torque; call this M, to avoid confusion with the rope tensions. Now because you calculate the truck acceleration numerically, you can use this to calculate the drum angular acceleration, and then the linear acceleration of m_2. So when you come to the algebra, you will have only three equations to solve.

87

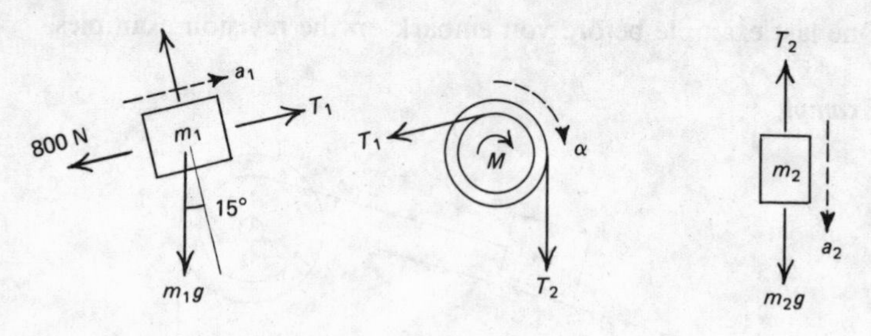

Kinematically, for m_1: $\quad x = ut + \frac{1}{2}at^2$

$$\therefore\ 10 = 0 + \tfrac{1}{2} \times a_1 \times 8^2$$

$$\therefore\ a_1 = 0.3125\ \text{m}\,\text{s}^{-2}$$

Drum: $\quad \alpha = \dfrac{a_1}{r_i} = \dfrac{0.3125}{0.24} = 1.3021\ \text{rad}\,\text{s}^{-2}$

Hanging weight m_2: $\quad a_2 = \alpha \times r_o = 1.3021 \times 0.32 = 0.4167\ \text{m}\,\text{s}^{-2}$

Now we know all the accelerations numerically, the equations of motion for the two masses can be solved completely—that is, T_1 can be found from the equation for m_1, and T_2 from the equation for m_2. (By the way, I hope you did not make the mistake of assuming the acceleration for m_2 to be the same as for m_1.) The equation of motion for the drum can then be written, and M calculated. The work is completed in the next frame.

88

The equations of motion are

For m_1: $$T_1 - 800 - m_1 g \sin 15^\circ = m_1 a_1 \quad (1)$$

$$\therefore T_1 = 1250 \times 0.3125 + 800 + 1250 \times 9.81 \times 0.2588 = 4364.16 \text{ N}$$

For m_2: $$m_2 g - T_2 = m_2 a \quad (2)$$

$$\therefore T_2 = 200 \times 9.81 - 200 \times 0.4167 = 1878.66 \text{ N}$$

For drum: $$M + T_2 \times 0.32 - T_1 \times 0.24 = I \times \alpha \quad (3)$$

$$\therefore M = 250 \times 1.3021 - 1878.66 \times 0.32 + 4364.16 \times 0.24 = \underline{771.8 \text{ N m}}$$

You can see from the component figures making up this value for M that the presence of the counter-weight means that very much less torque is required from the driving motor to haul the truck up the slope. Nearly all hoists make use of this principle. Most of the load is balanced by a counter-weight; the driving motor is required to overcome only a small difference of loads, togehter with any friction effects, and, of course, the extra forces needed to accelerate the various components of the system.

Right. Now have a go at the 'drill' exercises for this section in Frame 89.

89

'Drill' exercises: connected systems

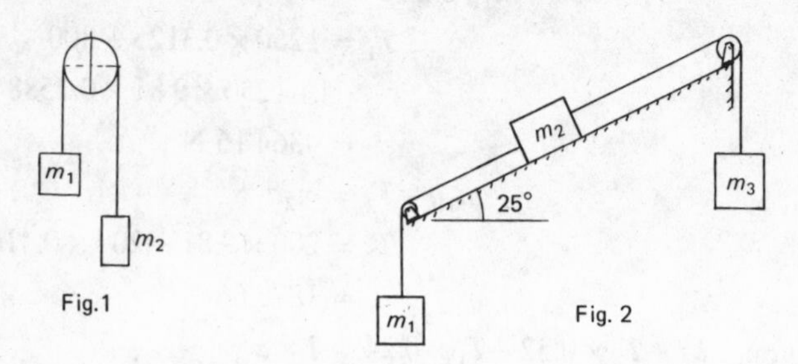

1. Bodies of masses m_1 and m_2 of magnitudes 3 kg and 4 kg, respectively, are connected by a light inextensible string which passes over a frictionless peg, as shown in Fig. 1 above. The system is released from rest, and exactly one second after release, a third body of mass $1\frac{1}{2}$ kg is added to m_1. Calculate how far upwards m_1 travels from its initial position of rest before coming to rest again. Calculate also the total time elapsed before it returns to its initial position.
Ans. 2.4022 m; 6.3136 s.
2. Bodies of masses m_1, m_2 and m_3, having respective magnitudes of 4 kg, 6 kg and 5 kg, are connected by two light inextensible strings passing over frictionless pulleys, as shown in Fig. 2 above. m_2 lies on an inclined plane of slope 25°. The system is released from a state of rest. Calculate the acceleration of the system (a) if the plane is smooth, (b) if the coefficient of friction between m_2 and the plane is 0.1, (c) if the coefficient of friction is 0.3.
Ans. (a) 1.004 m s^{-2} (m_1 downwards); (b) 0.6487 m s^{-2} (m_1 downwards); (c) 0.

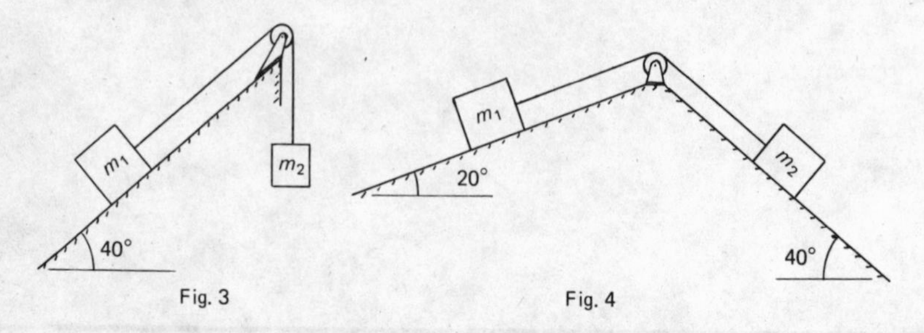

3. Bodies of masses m_1, 12 kg and m_2 10 kg are connected by a light string passing over a light frictionless pulley; m_1 lies on an inclined plane of slope 40° while m_2 hangs freely, as shown in Fig. 3 above. The friction coefficient between m_1 and the plane is 0.1. m_1 is given an initial downward velocity of 2 m/s. Calculate (a) the time for the system to come to rest, (b) the additional time for m_1 to reach its starting-point, (c) its velocity when it reaches its starting-point.
Ans. (a) 1.3991 s; (b) 2.1423 s; (c) 1.3061 m/s.
4. Figure 4 above shows two bodies, of masses, m_1 and m_2, 2 kg and 3 kg respectively, connected by a light inextensible string which passes over a light frictionless pulley. The bodies rest on inclined planes of slopes 20° and 40°. When the system is released from rest, m_1 is observed to move up the plane, and travels a distance of 2 m in 4.38 s. Assuming the coefficient of friction between body and plane to be the same for both planes, and assuming m_1 moves with a constant acceleration, calculate the coefficient of friction.
Ans. 0.272.

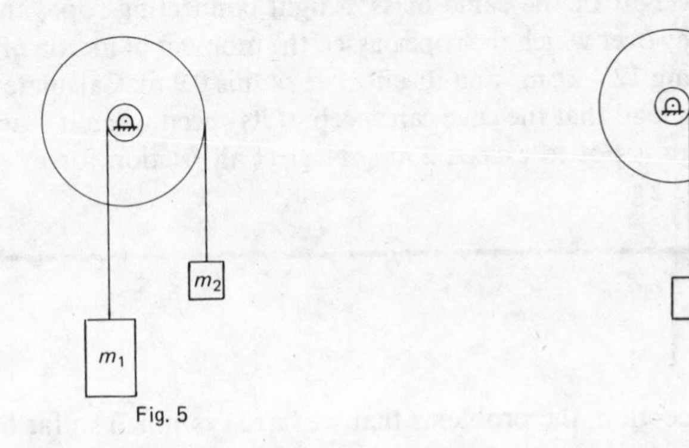

Fig. 5

Fig.6

5. Figure 5 above shows a simple wheel-and-axle type hoist. It consists of a wheel of radius 1.2 m mounted on an axle of radius 0.2 m, the moment of inertia of the whole assembly being 14 kg m². m_1 is the load which is to be lifted; this is attached to the axle. The weight of m_2, which is attached to the wheel, causes the load to be raised. Friction may be assumed to be negligible throughout. If load m_1 is 28 kg, calculate the value of m_2 sufficient to raise m_1 a distance of 10 m in 15 seconds.
Ans. 5.54 kg.

6. Figure 6 above shows a simple wheel and axle turning about a fixed axis. A body of mass 1 kg is attached to a light inextensible string which is wound round the axle which has a radius of 12 mm. The system is released from rest and the wheel begins to turn. The body is arranged to drop off after the wheel has completed exactly 12 turns, and this occurs 4.8 seconds after release. The wheel makes an additional 9 complete turns before coming to rest due to friction at the axle. Determine the moment of inertia of the wheel.
   Hints: assume a frictional torque F acting on the wheel all the time. Use kinematic formulae to evaluate the angular accelerations before the body falls off, and after it has fallen off. (*Ans.* 6.545 rad s^{-2} and 8.7266 rad s^{-2}.) Write equations of motion for wheel and body before, and for wheel after body falls off. Eliminate F from the equations.
   Ans. 0.00765 kg m^2.
7. A simple gravity-operated hoist is required to lower a load a vertical distance of 3.6 m. The hoist consists of a load-cage of mass 120 kg, a counter-weight of the same mass, a light connecting-rope, and a fixed pulley over which the rope passes, the moment of inertia of the pulley being 12.4 kg m^2 and its effective radius 0.9 m. Calculate the maximum load that the cage can accept if its speed when it reaches the bottom is not to exceed 2 m/s. Neglect all friction forces.
   Ans. 15.33 kg.

90

With one exception, the problems that we have examined so far have dealt with forces acting on bodies where the shape of the body is not relevant to the problem; so far as the problem has been concerned, we have been able to assume that the mass has been concentrated at a point. As we stated in Frame 6, the body was treated as a particle. The exception was the effect of forces acting on rotating bodies (Frames 53 to 66). We now need to extend our work to include problems of forces acting on bodies wherein the *shape* of the body is important. For example, in Frame 28 we looked at the forces acting on a car which was travelling around a banked circular track. We must now be prepared to look at this problem in more detail. Part of the solution consisted in finding the reaction force R between car and road; in this section, we shall learn to calculate how this reaction is shared between the inner and outer wheels of the car. In such cases, we drop the term 'particle' and refer instead to a 'rigid body'.

As a starting point we consider a rigid body subjected to a single force F; the line of action of this force does not pass through the mass centre, G, of the body.

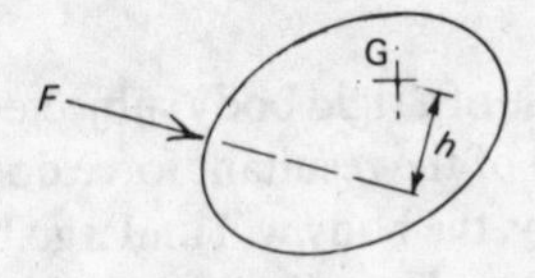

To illustrate more clearly the effects of this force, we first include two equal and opposite forces, F, parallel to the original force and passing through the mass centre G. Since the forces are equal and opposite, they cancel out and, dynamically, the body will behave as if subjected to the single original force.

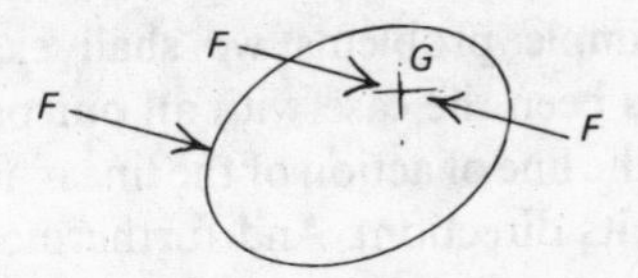

We now split this 'system' into two components

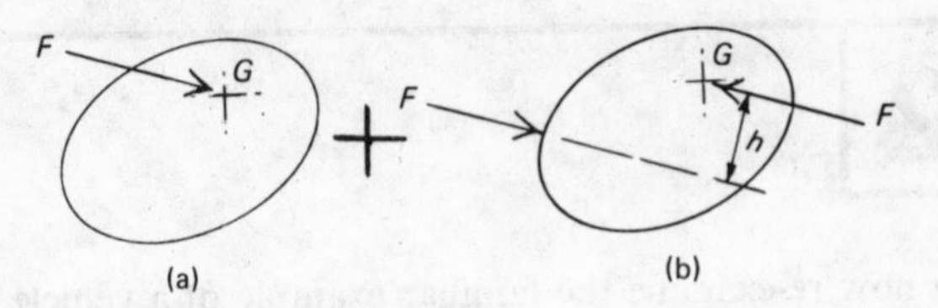

(a) comprises a single force F, parallel to the original force, but now passing through G. (b) comprises a *couple*, of magnitude $(F \times h)$ where h is the perpendicular distance from G to the line of action of the original F.

The single force F at G shown in (a) will produce a linear acceleration, a, of the body; this is given by

$$F = m \times a$$

The couple $(F \times h)$ shown in (b) will produce an angular acceleration, α, of the body; this is given by

$$F \times h = I_G \times \alpha$$

where I_G is the moment of inertia of the body with respect to an axis through G.

91

Thus, in the general case of a rigid body subjected to a number of forces, when the line of action of the resultant force does not pass through the mass centre of the body, the body will undergo both linear acceleration and angular acceleration. Here is a simple illustration. A small boat floats on a lake, and you attach a rope to it to pull it ashore. If the boat were originally broadside-on to you, and you fix the rope to the bow (or the stern), when you pull the rope, the boat, in addition to moving towards the shore, will also turn, and it will continue to turn until the line of the rope passes through the mass centre of the boat—that is, when the boat is more or less pointing towards you.

However, the simple problems we shall examine will be more constrained. As has been the case with all our previous problems, we shall always know the line of action of the linear acceleration (although we may not know its direction). And furthermore, we shall consider only cases where there will be no angular acceleration.

92

We may now re-examine the familiar example of a vehicle travelling along a road, but we shall now consider the vehicle as a *rigid body* and not as a particle, as we have done so far, in, for example, Frames 25 to 30. We shall draw a free-body diagram of, say, a car, taking into account its size and shape: when examining all the forces acting on the car, we have to take into account not only their direction, but also the points at which they act. Consider first a car being accelerated along a straight level track, and think of all the forces which must act on it. Here is a simple sketch of the car, shown as a car, instead of as a dot, or a rectangle.

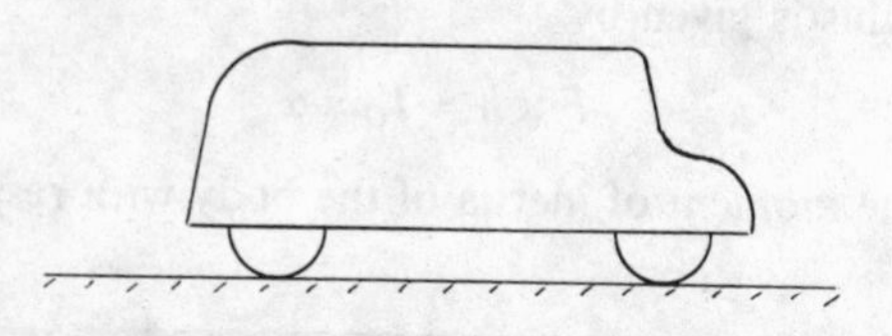

List all the forces you can think of, neglecting resistance of the air. Do not forget that the car is being accelerated; the engine must therefore apply a tractive force to the body of the car. Refer back to Frame 9. Including the tractive force, you should list four forces.

93

The forces comprise

1. The weight of the car. Since we are now concerned with the points of application of the forces, we shall require to know the location of the mass centre of the car.
2. The tractive force. If you clearly understood Frame 9, you will know that this force is exerted by the road, on the driving wheels of the car (in opposition to the thrust of the driving wheels on the road). The force must therefore act tangentially along the line of the road or track. It is irrelevant whether the car is driven by the rear wheels, the front wheels, or both front and rear; the line of action of the force will always be tangential to the wheels.
3. There will be a normal reaction component of road thrust on wheels, for *each* pair of wheels (that is, front and rear). Previously, we have considered road reaction as a single force only. These two reactions make up the total of four forces acting.

94

Before we attempt a numerical example, we must consider the effect of these four forces on the car. Although we have seen that, in general terms, the effects of a force system on a rigid body will be a combination of linear and angular acceleration, it should be realised that, apart from rather unusual and disastrous cases, the forces on a vehicle combine to produce linear acceleration only—forward or backward acceleration along the line of the track. Assuming this condition, we can conclude:

(a) that the resultant force on the vehicle passes through the mass centre
(b) that its line of action is parallel to the road or track
(c) and that, in consequence of (b), the resultant component perpendicular to the line of the track must be zero.

Conclusion (a) leads us to an even more significant conclusion: since the resultant force on the body passes through the mass centre, then the resultant *moment* of all forces about the mass centre must be zero. Be careful at this point to note that this condition applies to moments of forces *about the mass centre*. When a body is subjected to a resultant force through the mass centre, it should be clear that the moment of this force about any point which is *not* the mass centre need not be zero. In solving problems of forces on rigid bodies, students are prone to make the mistake of writing a moment equilibrium equation about a point other than the mass centre.

Now we will solve a simple problem.

Example A car has a mass of 850 kg. The distance between front and rear wheel centres is 2.7 m. The mass centre of the car is 1.2 m to the rear of the front wheel centre, and 0.9 m above the road level. Analyse the forces on the car when it accelerates forward along a straight level road at 2 m s^{-2}.

95

We have already considered the forces; here is the free-body diagram.

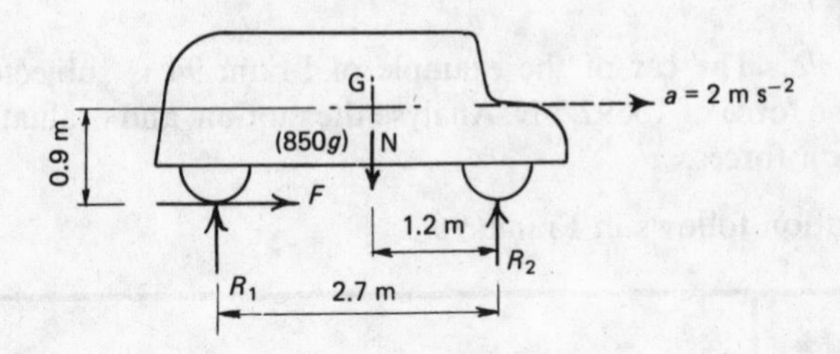

We do not need to resolve any of the forces. As with several earlier examples, we can write two equations

Along track: $(\Sigma F = m \times a)$ $\quad F = 850 \times 2 = 1700$ N $\qquad$ (1)

Across track: $(\Sigma F = 0)$ $\quad R_1 + R_2 = 850\,g$ $\qquad$ (2)

Unlike earlier examples, these two equations are insufficient to determine all forces. This is where we need the equation of moment equilibrium.

Moments about G: (assuming clockwise as positive)

$$R_1\,(2.7-1.2) - R_2 \times 1.2 - F \times 0.9 = 0 \qquad (3)$$

Substituting for F from equation (1) and for R_2 from equation (2):

$$R_1 \times 1.5 - (850\,g - R_1)1.2 - 1700 \times 0.9 = 0$$

$$\therefore R_1 = \frac{1700 \times 0.9 + 850 \times 9.81 \times 1.2}{(1.5+1.2)}$$

$$= \underline{4272.7 \text{ N}}$$

From equation (2):

$$R_2 = 850 \times 9.81 - 4272.7$$

$$= \underline{4065.8 \text{ N}}$$

Using the rules of simple statics, calculate what the values of front and rear wheel reaction forces would be if there were no acceleration (for example, if the car were standing still). You should find that R_1 would be 3706.0 N and R_2 4632.5 N. It can be seen that the effect of driving the car with a forward acceleration is to reduce the front-wheel reaction and increase the rear-wheel reaction. If you watch a car pull away from a standing start, you can see how the front lifts up. Similarly, when a car is braked suddenly, the front dips down.

As an interesting follow-up to this example, try and solve the next one yourself.

Example The car of the example of Frame 94 is subjected to a tractive force of 13897.5 N. Analyse the motion, and evaluate wheel reaction forces.

The solution follows in Frame 96.

96

You may by now have discovered the significance of the rather accurately prescribed value of F.

We can use the same free-body diagram as previously (Frame 95). The equations are

Along track ($\Sigma F = m \times a$):

$$13897.5 = 850 \times a \qquad (1)$$

$$\therefore\ a = \underline{16.35\ \mathrm{m\,s^{-2}}}$$

Across track ($\Sigma F = 0$): $\quad R_1 + R_2 = 850\,g = 8338.5 \qquad (2)$

G↻ $\quad F \times 0.9 + R_2 \times 1.2 = R_1(2.7 - 1.2) \qquad (3)$

Substituting for F and for R_2

$$13897.5 \times 0.9 + 1.2(8338.5 - R_1) = R_1 \times 1.5$$

$$\therefore\ R_1 = \frac{13897.5 \times 0.9 + 8338.5 \times 1.2}{2.7}$$

$$= \underline{8338.5\ \mathrm{N}}$$

$$\therefore\ \underline{R_2 = 0}$$

This particular value for F of 13897.5 can be seen to be the limiting value for the condition that the car stays on the road. The front-wheel reaction force is reduced to zero, and the rear-wheel force equals the weight of the car. An increase of F would increase the anti-clockwise turning moment about G. The clockwise turning moment is provided by R_1, and this cannot increase, because the value of R_1 cannot increase to more than the car's weight. So an increase of F above this figure would actually cause the car to turn over backwards—one of the disastrous results hinted at in Frame 94. Needless to add, this value of F is hypothetical; no ordinary vehicle is capable of providing a tractive force of such magnitude. There are examples, however, of certain high-power sporting vehicles which could turn over. It is partly for this reason that high-speed sports cars are designed with a very low centre of mass.

The problem can be extended to motion along a sloping track.

Example The car of the example in Frame 94 is driven up a 15° slope with a forward acceleration of 2 m s^{-2}. Determine the tractive force and the magnitudes of front-wheel and rear-wheel reaction forces.

Here is the free-body diagram.

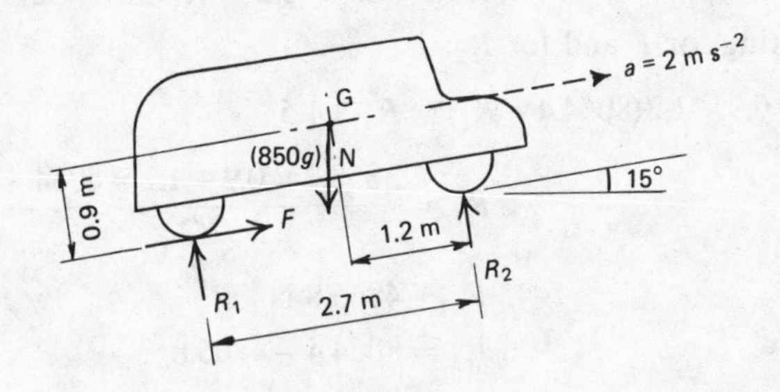

Attempt the three equations yourself before checking in Frame 97.

97

Observe that the weight, mg, must now be resolved in directions along, and perpendicular to, the track.

Along track ($\Sigma F = m \times a$): $F - mg \sin 15^\circ = m \times a$

$$\therefore F = 850 \times 2 + 850 \times 9.81 \times \sin 15^\circ = \underline{3858.2 \text{ N}}$$

Across track ($\Sigma F = 0$): $R_1 + R_2 = mg \cos 15^\circ$

$$= 850 \times 9.81 \times \cos 15^\circ = 8054.4 \text{ N}$$

G↻ $\quad F \times 0.9 + R_2 \times 1.2 = R_1 (2.7 - 1.2)$

Substituting for F and for R_2:

$$3858.2 \times 0.9 + 1.2(8054.4 - R_1) = R_1 \times 1.5$$

$$\therefore R_1 = \frac{3858.2 \times 0.9 + 1.2 \times 8054.4}{2.7}$$

$$\therefore R_1 = \underline{4865.8 \text{ N}}$$

$$\therefore R_2 = 8054.4 - 4865.8 = \underline{3188.6 \text{ N}}$$

Bearing in mind that applying brakes produces a negative 'tractive force', have a try at solving the next example yourself.

Example The car of the previous example is coasting down a 20° slope when the brakes are applied to produce a retardation of 2.8 m s^{-2}. Determine the braking force, and the magnitudes of front and rear wheel reaction forces.

The answers should be: $F = 5231.9$ N; $R_1 = 1738.5$ N; $R_2 = 6097.1$ N. The solution follows.

98

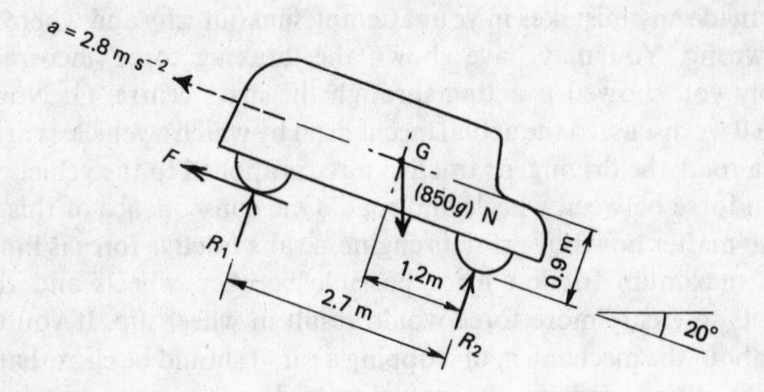

Since the car is being retarded, acceleration must be directed up the slope. The equations

Along track ($\Sigma F = m \times a$):

$$F - mg \sin 20^\circ = m \times a$$

$$F = m \times a + mg \sin 20^\circ$$

$$= 850 \times 2.8 + 850 \times 9.81 \sin 20^\circ$$

$$= \underline{5231.9 \text{ N}}$$

Across track ($\Sigma F = 0$):

$$R_1 + R_2 = 850\,g \cos 20^\circ$$

$$= \underline{7835.6 \text{ N}}$$

Moments about G:

$$F \times 0.9 + R_1(2.7 - 1.2) = R_2 \times 1.2$$

Substituting for F and R_2:

$$5231.9 \times 0.9 + R_1 \times 1.5 = 1.2(7835.6 - R_1)$$

$$R_1 = \frac{1.2 \times 7835.6 - 5231.9 \times 0.9}{2.7}$$

$$R_1 = \underline{1738.5 \text{ N}}$$

$$R_2 = 7835.6 - 1738.5 = \underline{6097.1 \text{ N}}$$

99

If you made any mistakes in your attempt, find out why and where you were wrong. You may have shown the 'braking force' incorrectly; possibly you showed it acting through the mass centre, G. Now in Frame 9 we discussed the actual mechanism by which a vehicle is driven along a road; the driving, or tractive force is applied to the vehicle as a friction force between wheels and road. One consequence of this was that, no matter how powerful its engine, a car's tractive force is limited to the maximum friction force possible between wheels and road; attempts to create more force would result in wheel slip. If you now think about the mechanism of stopping a car, it should be clear that the same conditions apply; the actual retarding force consists of a tangential friction force between wheels and road. So a car's braking system is only as effective as this maximum friction force; attempting to brake the car harder would result in the car, with wheels locked, sliding along the road. You can, of course, increase the friction force by increasing the normal reaction force between wheels and road, but you can do this only by adding to the mass of the car, with the result that the extra braking force would be used up in retarding the extra mass.

100

In Frames 25 to 30 we looked at problems of vehicles travelling around circular tracks, but we treated the vehicle as a particle. In Frame 28, the free-body diagram shows the reaction force R between wheels and road as a single force. Furthermore, all three forces (R, weight and friction, F) were treated as through they all passed through a single point. To round off this part of the text, we shall now return to this problem of a car on a circular track, this time treating the car as a rigid body. For this purpose, we shall require to know the distance between near-side and off-side wheels, and the location of the mass centre, G.

Example A car of mass 850 kg has a track width (that is, distance between near-side and off-side wheels) of 1.8 m. The mass centre is located centrally between wheels, at a height of 0.8 m above the ground. The car travels at constant speed of 24 m/s around a circular track of mean radius 45 m, and the track is banked towards the centre at an angle of 35°. Assuming that the car does not slip, calculate all the forces acting on the car under these conditions.

We shall draw a free-body diagram looking from the rear of the car, assuming that it is turning to the right. The forces will consist of

1. The weight; vertically downwards, acting at G.
2. A reaction force R_1 between outer wheels and road, acting upwards, its direction being perpendicular to the track surface. (Actually, this force will comprise two separate forces, on front and rear wheels, but we may treat these as a single force.)
3. A similar reaction force R_2 between inner wheels and road.
4. A friction force F between wheels and road, acting on the lowest point of the wheels, the line of action being parallel to the road surface. If the track were flat instead of being banked, the motion around the circle would result in a centripetal acceleration to the right; this must be provided by the sideways friction between wheels and road. We may therefore assume that the force acts from left to right—that is, down the slope of the banked track.

The resulting free-body diagram is therefore like this

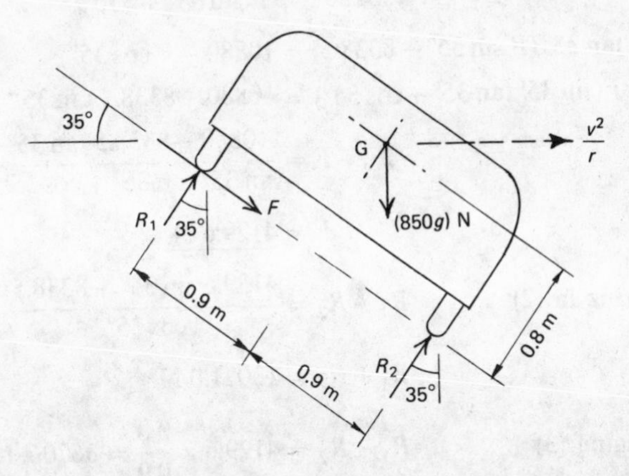

Notice, as with our earlier problems, that the direction of the centripetal acceleration is horizontal.

Attempt to write the three equations before reading on.

101

Here are the three equations.

Horiz. ($\Sigma F = m \times a$):

$$R_1 \sin 35° + R_2 \sin 35° + F \cos 35° = 850 \times \frac{24^2}{45} = 10880 \quad (1)$$

Vert. ($\Sigma F = 0$):

$$R_1 \cos 35° + R_2 \cos 35° = F \sin 35° + 850\,g \quad (2)$$

Moments about G: $$R_1 \times 0.9 = R_2 \times 0.9 + F \times 0.8 \quad (3)$$

(Check for yourself that the left-hand side of equation (3) is a clockwise moment about G whereas the right-hand side consists of the two anti-clockwise moments.) Re-arranging (1) and (2) and dividing one by the other:

$$\frac{(R_1 + R_2)\sin 35°}{(R_1 + R_2)\cos 35°} = \frac{10880 - F \cos 35°}{F \sin 35° + 850\,g}$$

$$\therefore \tan 35° (F \sin 35° + 8338.5) = 10880 - F \cos 35°$$

$$\therefore F(\sin 35° \tan 35° + \cos 35°) = 10880 - 8338.5 \tan 35°$$

$$\therefore F = \frac{10880 - 8338.5 \tan 35°}{\sin 35° \tan 35° + \cos 35°}$$

$$\therefore F = \underline{4129.6\ \text{N}}$$

Substituting in (2): $$R_1 + R_2 = \frac{4129.6 \sin 35° + 8338.5}{\cos 35°}$$

$$\therefore R_1 + R_2 = 13071.0\ \text{N}$$

Re-arranging (3): $$R_1 - R_2 = 4129.6 \times \frac{0.8}{0.9} = 3670.2\ \text{N}$$

Subtracting: $$2R_2 = 9400.8\ \text{N}$$

$$\therefore R_2 = \underline{4700.4\ \text{N}}$$

Substituting: $$R_1 = 13071.0 - 4700.4 = \underline{8370.6\ \text{N}}$$

102

To gain practice in this sort of calculation, repeat this example, but this time, assume a speed of 30 m/s instead of 24 m/s. The form of the calculation will be exactly as above, and so it is not given here, but you should find that R_1 is 12354.1 N, R_2 is 4227.2 N and F is 9142.8 N. You can see, therefore, that the effect of increasing the speed is to increase F, to increase the outer-wheel reaction force R_1, at the same time decreasing the inner-wheel reaction force R_2. It becomes clear that there must be a speed for which the value of R_2 is zero. See if you can calculate this speed. Draw a new free-body diagram, generally similar to the one in Frame 100, but without the force R_2. Remember when calculating that the value of the speed v is now not known. The value you should obtain is 61.6 m/s. Write the three equations, as we have done in the previous frame, modified to conform to the new conditions. Then manipulate the equations to eliminate F and R_1. A solution follows in Frame 103.

103

Here is the free-body diagram

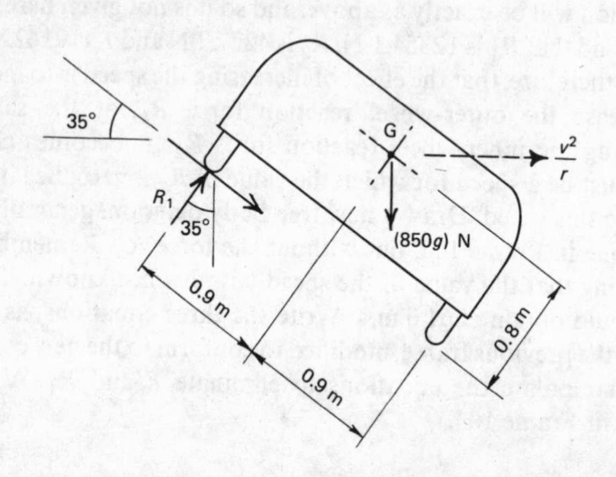

And the equations

Horiz. ($\Sigma F = m \times a$):

$$R_1 \sin 35° + F \cos 35° = 850 \times \frac{v^2}{45} \qquad (1)$$

Vert. ($\Sigma F = 0$): $$R_1 \cos 35° = F \sin 35° + 850\,g \qquad (2)$$

Moments about G: $$R_1 \times 0.9 = F \times 0.8 \qquad (3)$$

From (3): $$F = 1.125\,R_1$$

Substitute in (1) and (2), re-arranging

$$R_1 \sin 35° + 1.125\,R_1 \cos 35° = 850 \times \frac{v^2}{45}$$

$$R_1 \cos 35° - 1.125\,R_1 \sin 35° = 850 \times 9.81$$

Dividing, and cancelling R_1 and 850

$$\frac{\sin 35° + 1.125 \cos 35°}{\cos 35° - 1.125 \sin 35°} = \frac{v^2}{45 \times 9.81}$$

$$\therefore\ v^2 = 3795.9$$

$$\therefore\ v = \underline{61.6 \text{ m/s}}$$

We may, if we wish, substitute back and determine the corresponding values of F and R_1 but this is not required in the problem.

We need to explore the meaning of this example a little more deeply. We have found that at this speed of 61.6 m/s, the value of the inner-wheel reaction will be zero. What happens if we prescribe a speed of, say, 65 m/s? Well, you may do this if you wish. Repeat the calculation of Frame 101 yet again, using the free-body diagram of Frame 100, and, indeed, writing the same three equations, except that v is now 65 instead of 24. Your answers should be: $F = 60590.2$ N; $R_1 = 61771.7\ N$; $R_2 = -9166.6$ N. Interpreting this result, we conclude that if this vehicle is to travel around the circular path as prescribed, the inner wheels must be *held down* on to the track with this force R_2. If the physical conditions do not permit this negative downward reaction force (as, with an ordinary vehicle, they do not), then the vehicle cannot move as prescribed, and something else will happen. We must approach the solution, assuming no force at the inner wheels; we should then find that forces F and R_1 would combine to give a resultant anti-clockwise moment about G; in other words, the vehicle would tip over outwards, as we know would actually happen in fact.

104

Before leaving this type of problem, we must look at the force F. We established some distance back (Frame 26) that a car on a road requires friction between tyres and road to provide this sideways force. Banking the track means that we need rely less on friction to produce the centripetal acceleration, but in all our examples in this section, we have found that an inward force F was required. In every case, we have assumed that the force *can* be provided; this, of course, is no guarantee that it *will* be provided in a real situation. If the road were slippery, it might be impossible to provide a sideways force of the magnitude calculated, and again, the result would be that the vehicle would not move as assumed or prescribed. The force F must be limited by the friction conditions between road and wheels, and cannot exceed the value $\mu\ (R_1 + R_2)$, where μ is the friction coefficient. So, when vehicles travel around circular tracks, we find that there are actually two conditions limiting the motion; the first is the possibility of tipping over, and the second is the possibility of side-slipping.

If we refer back to the example solved in Frame 101, we obtained a value of 13071 N for $R_1 + R_2$. If we assume a value of 0.5 for μ the maximum possible corresponding value for F would be $\mu(R_1 + R_2) = 6535.5$ N. The calculated value of 4129.6 N is therefore within the allowable limit, and the car would not side-slip. If, however, you return to Frame 103, and calculate the values of R_1 and F, you will find that $R_1 = 47956.2$ N and $F = 53950.6$ N. For this to be possible, a friction coefficient of 1.125 is required—a most improbable value. So although we have shown that there is a limiting speed of 61.6 m/s beyond which the car must tip over, there is actually a second and lower value of limiting speed at which it would side-slip. This is not to say that a vehicle will always slip before it turns over; on a normal flat curve, tipping is usually much more likely than slipping. To round off this section, before going on to the 'drill' examples, calculate the limiting speed of the vehicle, assuming a maximum friction coefficient of 0.5. Use the same free-body diagram as in Frame 100, except that F must be replaced by $\mu(R_1 + R_2)$. v, of course, is not known, and is to be found.

105

The equations will be very similar to the solution of Frame 101.

Horiz. ($\Sigma F = m \times a$):

$$(R_1 + R_2) \sin 35° + 0.5(R_1 + R_2) \cos 35° = 850 \times \frac{v^2}{45} \quad (1)$$

Vert. ($\Sigma F = 0$):

$$(R_1 + R_2) \cos 35° - 0.5(R_1 + R_2) \text{in } 35° = 850\,g \quad (2)$$

In this case, a third equation is unnecessary. Dividing (1) by (2), and cancelling $(R_1 + R_2)$ and also 850

$$\frac{\sin 35° + 0.5 \cos 35°}{\cos 35° - 0.5 \sin 35°} = \frac{v^2}{45\,g}$$

$$\therefore\ v = \underline{28.55 \text{ m/s}}$$

This value, as we predicted, is higher than the 24 m/s assumed in Frame 101, but less than the limiting tipping speed of 61.6 m/s.

Now: DRILL!

106

'Drill' exercises: rigid bodies

1. A car has a mass of 3100 kg. The distance between front and rear wheel centres is 3.7 m; the mass centre is 1.3 m above road level and 1.1 m to the rear of the front-wheel centreline. Calculate the magnitudes of the rear and front wheel reaction forces when the car travels with a forward acceleration of 0.45 m s^{-2}: (a) up a slope of 18°; (b) down a slope of 20°. Calculate also what value of forward acceleration up the 18° slope would result in both front and rear wheel reaction forces being equal, and determine the corresponding value of tractive force, and wheel reaction force.
   Ans. (a) 12390.6 N; 16532.0 N. (b) 5331.5 N; 23245.5 N. 2.351 m s^{-2}; 16686.1 N; 14461.3 N. ((a): F = 10792.5 N; (b): F = −9006.2 N; that is, braking.)
2. A car has a mass of 1100 kg. The distance between front and rear wheels is 2.8 m and the mass centre is 0.9 m to rear of the front wheels, and 0.6 m above road level. If the coefficient of friction between road and wheels is 0.71, calculate the maximum forward acceleration possible (a) up a slope of $\sin^{-1} 0.1$, (b) down the same slope. For each case, determine the tractive force. Assume that only the rear wheels drive.
   Ans. (a) 1.646 m s^{-2}; F = 2890 N. (b) 3.608 m s^{-2}; F = 2890 N.
3. A car has the specification given in Problem 2. It travels at 25 m/s down a slope of 8°. The coefficient of friction between wheels and road is 0.65. Calculate the least distance in which it can be brought to rest. Assume effective braking on all four wheels. Calculate also the wheel reaction forces while braking.
   Ans. 63.14 m (a = 4.949 m s^{-2}), 1946.4 N, 8739.6 N.
4. A car has a mass of 1250 kg. The distance between inner and outer wheels is 2.3 m and the mass centre is symmetrically between the wheels and 0.85 m above road level. The car travels at a constant speed of 20 m/s round a circular track of mean radius 45 m. Calculate the total outer and inner wheel reaction forces, and the sideways friction force on the car: (a) if the track is flat; (b) if it is banked inwards at 25° to the horizontal.
   Ans. (a) 10237.5 N, 2025.0 N; 11111.1 N. (b) 9711.0 N, 6093.8 N; 4887.7 N.
5. A car has a mass of 920 kg. The distance between inner and outer wheels is 2.1 m and the mass centre is centrally between the wheels

and 0.75 m above ground level. The car travels at constant speed around a circular track of mean radius 32m and banked at 20°.

(a) Calculate the speed such that the inner wheel reaction force would be zero (that is, the car is about to tip outwards), assuming that the car does not slip.

(b) Calculate the speed such that the car begins to side-slip, assuming a friction coefficient between road and wheels of 0.62.

Ans. (a) 33.60 m/s. (b) 19.97 m/s.

6. A block rests on the floor of a vehicle which is ascending a hill of slope 15°. The vehicle starts from rest and accelerates uniformly, reaching 15 m/s in 50 seconds. Under these conditions, the block just begins to slide backwards along the floor of the vehicle. Calculate what retardation would now be required on the vehicle in order to cause the block to slip forwards along the floor of the vehicle.

   Ans. 5.378 m s^{-2} ($\mu = 0.3$).

7. The figure shows a truck of 1550 kg mass, with some principal dimensions. It is pulled up a 20° slope by a cable attached at a point P which is at height h above the ground; the line of action of the cable is parallel to the ground. The force in the cable causes a forward acceleration of the truck of 2.1 m s^{-2}. Calculate the force F. If h is 0.8 m, calculate the front and rear wheel reaction forces. Also determine what value of h would cause both front and rear wheel reaction forces to be equal. Neglect all friction forces.

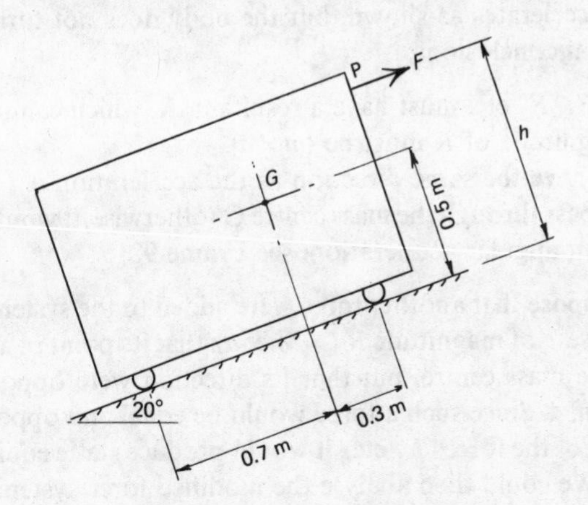

Ans. 8455.6 N; 1749.9 N. 12538.6 N; 0.162 m.

107

There is another approach to problems of kinetics, quite different from everything we have so far looked at.

Imagine a body of mass m, subjected to a number of forces, F_1, F_2, etc., as a result of which, it has an acceleration a in a certain direction. Thus

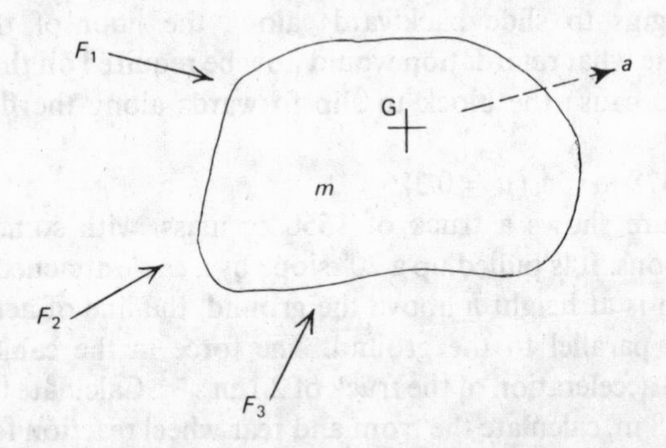

Furthermore, assume that there is no angular acceleration; the mass centre G accelerates as shown, but the body does not turn. We may draw several conclusions.

1. Forces F_1, F_2 etc. must have a resultant, R, which cannot be zero.
2. The magnitude of R must be $(m \times a)$.
3. R must have the same direction as the acceleration a.
4. R must pass through the mass centre G (otherwise, it would also give rise to an angular acceleration: see Frame 90).

Now suppose that another force were added to the system: that this new force were of magnitude R ($= m \times a$): that its point of application were G, the mass centre, but that its direction were opposite to the acceleration, a. Since such a force would be equal and opposite to the resultant R of the forces F_1 etc., it would produce static equilibrium of the body. We could then analyse the modified force system using the methods of *statics*; we would not need to resort to using the equation $\Sigma F = m \times a$, which we have had to do in all calculations so far in this programme.

This principle of changing a 'dynamics' problem into a 'statics' problem by adding an extra force is known as *D'Alembert's principle*.

We may summarise the procedure

1. Draw the free-body diagram, showing all real forces known to be acting on the body, and also the acceleration (so far as it is known, or can be assumed).
2. Include a '*reversed effective force*' on the diagram, of magnitude ($m \times a$) and direction opposite to that of the acceleration a. When the body has no angular acceleration, the point of application of the force is the mass centre, G.
3. Write sufficient equations of static equilibrium to solve for the unknown quantities.

You should recall the principles of static equilibrium:

(1) The resultant of all forces in any direction = 0

(2) The sum of the moments of all forces about *any* point = 0

The opening statement of this frame must be repeated. This is another approach to problems that you have hopefully, by now, learned already to solve. The reason for its inclusion in this problem is that you have very probably encountered it previously, and may have been puzzled that the work of the programme so far conflicts with what you 'learned' previously. The quote-marks are intended to suggest that although you may have been able to obtain correct answers to a few problems, you probably did not really understand exactly what you were doing!

108

A single example will be sufficient to illustrate the technique. We shall take the example of Frame 97. To save looking back, here is the example, restated.

> *Example* A car has a mass of 850 kg. The distance between front and rear wheel centres is 2.7 m. The mass centre is 1.2 m to the rear of the front wheel centre, and 0.9 m above road level. The car is coasting down a slope of 20° when the brakes are applied to produce a retardation of 2.8 m s^{-2}. Determine the braking force, and the magnitudes of the front and rear wheel reaction forces.

We draw the free-body diagram. This will be identical to that of Frame 98 but must now also include an extra 'force', acting at G, in the forward direction (because the acceleration is backwards) and of magnitude $(m \times a) = 850 \times 2.8 = 2380$ N. The word 'force' is in quote-marks because there is, of course, no real force acting here at all. It is a fictional force, included merely to solve the problem. For this reason, it is good technique to show it distinctly on the free-body diagram, so that it cannot be confused with the real forces, F, R_1, R_2 and mg.

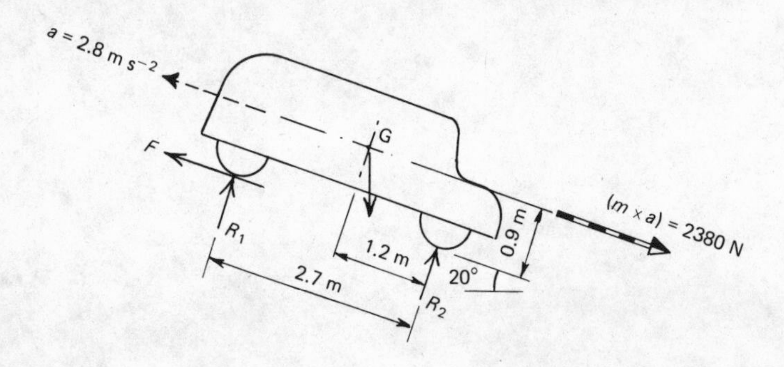

An equation of static equilibrium down the slope will yield the value of F

$$2380 + 850\,g \sin 20° - F = 0$$

giving: $$F = \underline{5231.9 \text{ N}}$$

Writing equilibrium of moments about the point of contact between front wheel and road

$$R_1 \times 2.7 + 2380 \times 0.9 = 850\,g \cos 20^\circ \times 1.2 - 850\,g \sin 20^\circ \times 0.9$$

giving: $\underline{R_1 = 1738.5 \text{ N}}$

Equilibrium across the slope

$$R_1 + R_2 = 850\,g \cos 20^\circ = 7835.6 \text{ N}$$

giving: $\underline{R_2 = 6097.1 \text{ N}}$

(Alternatively, R_1 can be determined by a moment equation about the rear-wheel contact point.)

The solution is simpler than that of Frame 98; there, we had to solve two simultaneous equations in R_1 and R_2. Here, we have found each unknown from a single equation. Recall that in Frame 99, you were warned that you could write a moment equilibrium equation only with respect to G, the mass centre. Here, because we are solving a 'static' problem, we may write a moment equilibrium equation with respect to *any* convenient point.

109

This brings us to the end of a rather long programme. There are no 'drill' examples for this final section. Remember, the use of D'Alembert's principle is an alternative method of solving kinetic problems, so if you would like some practice in using the technique, make use of the examples of Frame 106.

For those problems involving a vehicle travelling round a circular track, you know that the centripetal acceleration is v^2/r, and that it is directed towards the centre of the curve. Your 'reversed effective force' will therefore be mv^2/r, and will be directed radially outwards. Never forget that this 'force' is not a real force, but an imaginary or fictitious one. Most certainly, it is *not* the centrifugal force. Centrifugal force is a real force; not a fictitious one. In the case of a car on a circular track, the centrifugal force is the radially outward force exerted *by* the wheels of the car *on* the road. As such, it is not included in the system of forces shown on the free-body diagram, which purports to show only those forces acting *on* the body in question, and not the forces exerted by it on other bodies or constraints.

It is usually when analysing the kinetics of bodies having circular motion that students are first introduced to this concept of a radially outward force of magnitude mv^2/r, usually when they are perhaps too young and inexperienced to appreciate the rather sophisticated concept of making use of a fictitious force to translate a dynamics problem into a statics one. It must be made clear that even at this stage, it is not necessary for you to learn the technique. But if you have been introduced to the method earlier, without clearly understanding what you have been doing, it is certainly necessary for you to tidy up your ideas, and clarify your thinking. Later, in more advanced work, you may encounter problems wherein the use of D'Alembert's principle becomes almost essential. But in elementary work, you can easily manage perfectly well without it.

Index